PRAISE FOR *RACHEL CARSON AND HER SISTERS*

"A vibrant, engaging account of the women who preceded and followed Rachel Carson's efforts to promote environmental and human health. In exquisite detail, Musil narrates the brilliant careers and efforts of pioneering women from the 1850s onward to preserve nature and maintain a healthy environment. Anyone interested in women naturalists, activists, and feminist environmental history will welcome this compelling, beautifully written book."—Carolyn Merchant, professor of environmental history, philosophy, and ethics, University of California, Berkeley, and the author of *The Death of Nature*

"With deep grounding in women's history, environmentalism, and public health—and, just as importantly, with great reverence—Musil introduces us to a pantheon of remarkable women, true heroines every one. This book offers a new perspective, countless wonderful stories, and inspiration. A great read!"—Howard Frumkin, MD, DrPH, dean, University of Washington School of Public Health; former director, CDC, National Center for Environmental Health

"Bob Musil brilliantly documents the rich trajectory of women's intellectual and political influence, not just on environmentalism but on public policy and activism. Musil offers fascinating details of Rachel Carson's struggles to be taken seriously as a scientist and unearths the stories of the women—unsung heroes all—who influenced her. A must read for anyone interested in American history, science and environmental politics."—Heather White, executive director, the Environmental Working Group

"An absolutely wonderful book! Bob Musil shows Rachel Carson not as a lone voice, but an eloquent one who drew inspiration from female predecessors and those around her. He argues persuasively that we can understand Carson better if we see her in relation to other women, to the broader environmental movement, and to working in community. Should be required reading for anyone interested in where we have been, and where we need to go." —Geoffrey Chase, dean, San Diego State University, and the author of *Sustainability in Higher Education*

"An eloquent and moving tribute to the women at the heart and soul of the environmental movement. It is a story of brilliant science, courage, stamina, and a passion for life. We are in debt beyond counting to them and to Robert Musil for telling their stories so well." —David W. Orr, Paul Sears Distinguished Professor of Environmental Studies and Politics, Oberlin College

RACHEL CARSON AND HER SISTERS

RACHEL CARSON AND HER SISTERS

Extraordinary Women Who Have
Shaped America's Environment

ROBERT K. MUSIL

RUTGERS UNIVERSITY PRESS
New Brunswick, New Jersey, and London

Second cloth printing, 2014

Library of Congress Cataloging in Publication Data
Musil, Robert K., 1943–
Rachel Carson and her sisters : extraordinary women who have shaped America's environment
   p.   cm.
Includes bibliographical references and index
ISBN 978-0-8135-6242-1 (hardcover : alk. paper) —ISBN 978-0-8135-6243-8 (e-book)
1. Carson, Rachel, 1907–1964. 2. Women biologists—United States—Biography. 3. Women environmentalists—United States—Biography. 4. Science Writers—United States—Biography.
QH31.C33M87 2014
508.092—dc23
[B]

                                                                        2013021944

A British Cataloging-in-Publication record for this book is available from the British Library.

Visit our website: http://rutgerspress.rutgers.edu

Manufactured in the United States of America

To my wife, Caryn McTighe Musil

An exemplary national leader in women's studies
and higher education

My loving partner in all things personal and political since
the days of miniskirts and mobilizations

CONTENTS

FIGURES

PREFACE AND ACKNOWLEDGMENTS

Rachel Carson sits alone on a rock outcrop atop the Hawk Mountain Sanctuary with her binoculars. It is a striking photo from 1945 that captures the most visionary environmental scientist, writer, and advocate of the twentieth century. It is an Olympian pose, far above the fields and foliage and farms of Pennsylvania below. In fewer than twenty years, Carson would be again alone, seated, testifying at U.S. Senate hearings on DDT and other deadly pesticides that were killing robins, eagles, and osprey, and that were harming humans, too. In 1963, as a best-selling nature writer, she was greeted by senators, as had been Harriet Beecher Stowe, author of *Uncle Tom's Cabin*, by President Abraham Lincoln, as "the little lady who started it all."

It is a lovely tribute, as is the photograph of her looking for migrating hawks and eagles. It invites us to scan the horizon, to look forward, to observe nature carefully, to note what it bodes for our future. But like many photos, and accolades as well, this one is as important for what it leaves out, for what is nearby, unseen, omitted. Carson was not alone on the mountaintop. This image of her was taken by a woman, Rachel's close friend and colleague, Shirley Briggs. Briggs was one of her fun, bright, underrated pals at the U.S. Bureau of Wildlife and Fisheries, where both had already become aware of the emerging evidence of the dangers of DDT. When Carson died of breast cancer on April 14, 1964, her work not yet done; Shirley Briggs carried it forward. She wrote an important book on pesticides and founded what is now the Rachel Carson Council.

The Hawk Mountain Sanctuary itself was fought for and founded by another woman, Rosalie Edge, a feisty champion of conservation and an unlikely close ally of Franklin Delano Roosevelt's confidante, the environmentalist secretary of the interior, Harold Ickes. Had it not been for Rosalie Edge, a photo taken a little more than a decade before would have shown men with guns slaughtering many of the twenty thousand or so raptors that migrate by this ridge each fall. Carson is using binoculars, instead of a gun, because another woman, Florence Merriam Bailey, the nation's first campus environmental organizer at Smith College in the 1880s, had written *Birds through an Opera Glass* (1889), which set off the craze of birding with binoculars.

Rachel Carson was the visible peak of a whole range of women who came before her and who carry on today. I simply try to make these others more noticeable, to remind us and me that women have been central to the environmental movement and our understanding of ecology since its earliest stirrings and fragile beginnings in the nineteenth century. In tracing these women's lives, accomplishments, and accomplices, I have learned much, been humbled, awed, and accrued a number of debts that I want to acknowledge here. As an avid birder, an aspiring author, and a long-time advocate for the environment and health as the former head of Physicians for Social Responsibility, I offer this book itself as an acknowledgment of the pioneering work of the women who fill these pages. They should be known to every American who cares about our nation, our children, our world.

I am indebted as well to the founders of women's history and to the contemporary women scholars, biographers, archivists, and activists who have kept alive the records and the realities of this long line of true American heroes. Linda Lear's monumental biography, *Rachel Carson: Witness for Nature* (1998), gave me inspiration, information, and my earliest intimations that Carson did not reach the pinnacle alone, but was the product of predecessors, and that she was supported throughout by steadfast female friends and a far-flung network of collaborators. Carolyn Merchant, Glenda Riley, Deborah Strom, Madelyn Holmes, Marcia Myers Bonta, Maxine Benson, Harriet Kofalk, Mary Joy Breton, Rochelle Johnson, Susan Goodman, and many, many others who are cited in the notes to this volume did the first research on many of the women I describe. Stephanie Fabritius of Centre College encouraged my work while I was in residence there, told me amazing stories about Fran Hamerstrom, and gave me signed copies of her books. There are some sympathetic men as well. I have used the works of Mark Lytle, Daniel Patterson, Carl Dawson, William Souder, and others who give environmental women their due. Peter Mickulas, an excellent environmental historian and my editor at Rutgers University Press, sought me out when he came to the press and has remained enthusiastic and encouraging at every step. Wade Greene of Rockefeller Financial Services has supported my work for decades with grants that allow me to write, teach, and advocate as I see fit. Jim Thurber of the Center for Congressional and Presidential Studies at American University has graciously given me an academic home, the chance to teach about environmental politics, and the freedom to call things as I see them. Wilson Peden, an editor at the Association of American Colleges and Universities, rescued me and assisted me with modern formatting and digital issues that the younger generation happily can take for granted. Beth Gianfagna of Log House Editorial Services, a top-notch copy editor, provided a sharp eye and very pleasant punctiliousness. For generous assistance and help with archives and photo acquisition and permissions, I happily acknowledge Ross Feldner, Monroe Meyersburg, and Dr. Diana Post of the Rachel Carson Council; Ben Panciera of the Linda Lear Center for Special Collections and Archives at Connecticut College; Debby Richards and Susan Baker at the Sophia Smith Archives at the Smith College Library; Malgosia Myc at the Bentley Historical Library of the University of Michigan; Deborah Douglas and Rob Doane of the MIT Museum; Rachel Rohrbaugh at the Rachel Carson Collection at the Chatham University Library; Joey Fones at the American University Library; and the staff at the Beinecke Rare Book Library and Rachel Carson Papers at Yale University. I also want to thank Monette Clark, the literary assistant to Terry Tempest Williams, and Terry herself; Sandra Steingraber; Devra Davis; and Theo Colburn, true sisters of Rachel Carson, for kindnesses and trailblazing work over many, many years. Thanks and apologies, too, to my friends and colleagues in the environment movement who are mentioned in the epilogue. These women deserve another whole set of books and wider recognition. I suspect, I hope, as Rachel Carson peered through her binoculars at the far-off hills from atop Hawk Mountain, that she could see them coming.

RACHEL CARSON AND HER SISTERS

INTRODUCTION

I HOPE YOU ENJOY READING *Rachel Carson and Her Sisters* as much as I did writing it. Like many Americans, I had heard of Rachel Carson and knew in my youth, vaguely, that her book *Silent Spring* (1962) had something to do with saving birds and reducing the use of pesticides. Then, at the time of her one-hundredth birthday, in 2007, I became intrigued that talk show hosts like Glenn Beck and Rush Limbaugh were launching fiery, fresh attacks on her more than forty years after she had died of breast cancer on April 14, 1964. I wanted to know more about the roots of such venom. I opened a book here, a website there. I read Linda Lear's wonderful biography, *Rachel Carson: Witness for Nature*. Then I actually read *Silent Spring*, cover to cover. One book led on to another. Before long, I realized that there had been other Rachel Carsons long before she was born, and that many women have built on her legacy since her untimely death. These extraordinary women have also been wonderful scientists, popular writers for large audiences, and ardent advocates for the environment and human health. They, too, have faced vicious attacks from industry and its allies in the establishment. They, too, faced bias because they were women.

But for more than a century and a half, they have persevered against incredible odds, created mutual support networks, and slowly built an environmental movement strong enough to draw opposition from those who profit from pollution. Along the way they wrote marvelous books that touched millions of Americans even as they peered into microscopes or through binoculars and passed around petitions for the policymakers of the day. And I learned to admire, even catch my breath in awe at the courage of these women who have shaped how we know and love the American environment.

I have written this book for readers like you. You have heard something about Rachel Carson and want to know more. Or, you were attracted to the idea that there are many more women environmentalists than you and I ever knew about. Or, like me, you enjoy the outdoors and nature and are something of an amateur naturalist, gardener, or active birder. Ideally, for my purposes, you may also belong to a reading group or book club, or some organized group like a garden club or the Audubon Society.

As you read on, you will discover that Rachel Carson is rightly credited with jump-starting the modern environmental movement back in the days of John F. Kennedy. U.S. senators called her the Harriet Beecher Stowe of that time, echoing Abraham Lincoln, who said of Stowe when he met her during the Civil War, "So you are the little lady who started it all." But as I read, I soon learned, as I suspected, that Rachel Carson was not alone. It takes more than a single book or person, no matter how powerful, to ignite a movement. Rachel Carson did not suddenly awaken a somewhat slumbering conservation movement, save the robins and peregrine falcons, or ban DDT all by herself. She had help. Lots of it. And then I discovered that she was not the first female environmentalist leader, or the only one who should be named within the pantheon of American male environmental heroes—John James Audubon, Henry David Thoreau, John Muir, John Burroughs, Theodore Roosevelt, David Brower, and so on. Nor is Rachel Carson the last extraordinary woman to inspire Americans who care about the environment. Women are writing today's *Silent Springs* and acting on Carson's principles even as you read this.

As I became more and more intrigued and inspired by Rachel Carson, I noticed that she, too, was an avid reader. She was taught to love books, writing, and nature by her mother, Maria McLean Carson. Maria Carson was a highly educated woman of the late nineteenth and early twentieth centuries who had to give up teaching, as was the rule for women then, when she married. She introduced Rachel to the nature study movement, which was led by women, and also to *St. Nicholas* magazine, where many female naturalists and popular writers published. Then I was hooked. I discovered and became entranced by the daughter's predecessors, the women that she and her mother read, or heard about. And, then, I just read backward in time, learning about earlier extraordinary women naturalists, scientists, and authors who had influenced those who influenced Maria and Rachel Carson. The women I describe are, like Carson, only the most conspicuous among a broad range of women involved with shaping our views of the American environment. You will meet a surprising number of local and regional female American environmentalists along the way, as well as their organizations, clubs, and advocacy groups. But I have chosen to focus on women who were prominent, widely read, or acknowledged in their day. To write a complete environmental history of women in the United States would take multiple volumes.

Rachel Carson and Her Sisters reflects my own happy journey of discovering extraordinary women. The first three chapters are full of women who deserve definitive biographies, reissued books, and a place in that pantheon of American environmental heroes. The women you will meet originally inspired the nation to love nature and birds—especially robins, whose agonizing deaths from DDT set off *Silent Spring.* They sought to protect them and other species and to set aside and preserve wilderness and natural parks. Rachel Carson's foremothers educated generations through books for youths and adults, founded natural history museums, and wrote astonishing amounts despite the diversions of motherhood, families, and gender bias. Once I passed through some magical closet door into the nineteenth century and earlier, I discovered, literally, a library full of interesting and important pioneering women environmentalists. I have focused mainly on those who foreshadow Rachel Carson's passions of amateur ornithology, scientific concern with pollution and environmental health, a love of language and writing, and active advocacy. But I have also been amazed and delighted by the many women I encountered in my research whom you will not find here—in early colonial agriculture and gardening; in botany, entomology, mountaineering, hiking, and seaside studies; in the painting and early photography of birds, shells, and ferns; and in scientific and medical education, environmental and civic organizations, and much more.

The second half of the book introduces you to modern women—active today. Again, I focus on those who most closely reflect Rachel Carson's wonder and witness for nature—through brilliant writing mostly for educated, general readers. Each shares a broad ecological concern for nature and us humans who are an integral part of it; each uses her writing and her reputation to advocate for policy and political changes, as did Carson. You will not find here, as a result, important women writers and environmental advocates like Barbara Kingsolver, who is known primarily as a novelist, or philosophical nature writers like Annie Dillard, or women known more for their activism, even though they write books, like Lois Gibbs. I also stopped short of including some of the women who are major contemporary figures and leaders of today. It would have required far too many more chapters, or, rather, another entire book, to explain and introduce the women who will someday be seen as shaping the American environment.

At the center of this book stands Rachel Carson herself, whose work reaches back toward the nineteenth century and forward into our own time. Linda Lear's groundbreaking biography is a trove of material for those of us who want to understand Carson as a person, as well as a writer and advocate. William Souder's recent biography, *On a Farther Shore*, which appeared as I was finishing my writing, stresses, as I do, that Rachel Carson combined concern with radioactive fallout from nuclear weapons tests along with the rain of pesticides from the air. Mark Lytle's *The Gentle Subversive* is closest to my view of Carson as a politically engaged activist, as well as scientist and writer.

But my goal is not only to restore or refresh our views of Rachel Carson, but to place her achievement within the context of a long line of women—activists, novelists, scientific writers, naturalists, and educators—without whom she would not have flourished, nor would have today's large environmental movement. I also try to highlight a wonderful network of women who supported and collaborated with Carson during her lifetime. Though less well known, her friends and colleagues like Shirley Briggs, Marie Rodell, Marjorie Spock, and others deserve some time in the spotlight.

Perhaps this is the place to admit that I am also an environmental health advocate. If I have a bias, it is in trying to help humanity and the environment of which we are a part and to reach out to the unconvinced or unconcerned. So, in fact, I know, admire, and have advocated alongside each of the main modern women I present in the second half of the book: Terry Tempest Williams, Sandra Steingraber, Devra Davis, and Theo Colborn. I could have included many, many more remarkable contemporary women, but these four best combine many of Carson's qualities of accurate science, evocative prose, appreciation by a wide readership, and active advocacy.

I am proud to be just a small part of these and other women's networks. I like to believe that I occasionally resemble the supportive and sympathetic men you will also meet, mostly in the background, of *Rachel Carson and Her Sisters*. Of course, highlighting women who are established, published authors with scientific training creates its own bias. I am aware of numerous, younger, more diverse women writers, activists, and advocates who are worthy of accolades, like Majora Carter or Janisse Ray. Their turn will come. I also do not wish to repeat the misapprehensions that I am trying to dispel about Rachel Carson and the environmental movement. Such progress that we have made was brought about by social movements filled with and led by little-known and never-to-be-known women. They sustained broad, supportive women's networks and female friendships as they faced gender bias and professional obstacles throughout their lives. It is they who have created the powerful waves of interest in the environment and other reforms that better-known writers and advocates, let alone presidents, have ridden.

Although I currently teach at American University and for years led a major environmental health organization, Physicians for Social Responsibility, I particularly draw inspiration from the women like Rachel Carson and her sisters who wrote for broad, even popular audiences. Many combined scientific training with real gifts of language, even poetic sensibility. They have successfully bridged the sciences and the humanities, using anecdote, stories, metaphors, and real people. They even anthropomorphized the little creatures they were trying to save. I hope to accomplish something similar in *Rachel Carson and Her Sisters*, to keep you engaged, and to give you a few "Wow! I didn't know that!" moments. There are also plenty of endnotes and references for those of you in interdisciplinary classes, or who simply want to seek out more books and information about the women you discover here. A few more things you should be aware of before you hit the "Buy Now" button or head to the cashier. While the women in *Rachel Carson and Her Sisters* wrote for general readers, they also urged their readers to take action; they all belonged to the leading social and political movements, especially environmental ones, of their time.

A quick caution. There is science inside these pages. But be assured it is closer to what my scientific colleagues think of as introductory science for nonscientists or nonscience majors. As you meet and learn more about Rachel Carson and others, you will encounter the science behind air and water pollution, endocrine disruption, cancer, radiation, and toxic chemicals. But I approach these topics mainly through the stories and lives of women authors who wanted desperately for you and me to not be put off by "pure" science—statistics, data, formulas, and lengthy Latinate terms. Instead, they wanted you to be led toward science by the imagination, empathy, the wonders of the natural world, and of words.

Like many of the women I admire and present, I have been trained in both the humanities and the sciences. This is why I present two traditions that Rachel Carson first successfully combined into modern ecology for mass audiences. There are naturalists and conservationists on the one hand—who focus more on preserving warblers or woodlands. There are environmental health scientists on the other. They worry more about the effects of pollution on people—not just penguins or peregrine falcons. I believe deeply that we readers and citizens need to understand each of these traditions and combine them, as Rachel Carson did so effectively, if we are to become engaged members of society and build a more powerful public movement with any hope of saving our own species and the earth.

Even after all the assaults on our ecosystems and on environmentalists themselves in the past couple of decades, I have faith that many of you who read this book will decide to become even more engaged, informed, active citizens. When a few years ago I wrote the opening sentence for my first book, *Hope for a Heated Planet*, I said simply, "This book is an act of faith." *Rachel Carson and Her Sisters* reflects that sentiment again. It is similar to that of writer Terry Tempest Williams and her colleagues who are trying to save the Utah wilderness. They say about the process of pouring your heart and soul and beliefs into hundreds of printed pages in hopes of finding a reader to educate, inspire, or move to action: "Writing is an act of faith."

In *Rachel Carson and Her Sisters*, I begin with nearly forgotten figures like Susan Fenimore Cooper, whose *Rural Hours* was America's first popular nature book, a big best seller in 1850 that influenced Thoreau years before he wrote *Walden*; Anna Botsford Comstock, whose 1911 *Handbook of Nature Study* capped a nationwide movement and influenced Maria Carson; Dr. Alice Hamilton who worked with Jane Addams of Hull House among poor urban workers with dread diseases. Hamilton went on to pass laws to protect them and became the first female professor at Harvard. Moving from the women who preceded Rachel Carson and whose wide-reaching connections and influence captivated me, *Rachel Carson and Her Sisters* looks at Carson herself and the women who mentored, encouraged, and worked alongside her. Take Marjorie Spock, the sister of the famed pediatrician and pioneer antinuclear activist Benjamin Spock. Far ahead of her time in the 1950s, Marjorie and her partner, Polly Richards, were among the first to bring suit, in 1957–1958, over the aerial DDT spraying of their organic garden in Brookville on Long Island's wealthy North Shore.

Our story then moves to the current generation of women inspired by Carson, or following in her footsteps: Terry Tempest Williams says directly, "Rachel Carson is a hero of mine." Sandra Steingraber is, like Carson, at once writer, scientist, and a woman with cancer. Her first book, *Living Downstream*, has become a classic that mixes poetic prose with passionate revelations of the links between environmental pollution and human health. She says, "Rachel Carson was my call to arms." Or take Devra Davis, a leader in linking the causes of cancer to the environment. Davis grew up in the polluted steel town of Donora, Pennsylvania, and watched her beloved grandmother, who escaped the Holocaust, die from industrial emissions. Her first book, *When Smoke Ran Like Water*, nominated for the National Book Award, mixes the personal, political, and scientific in the tradition of Rachel Carson. Throughout her career, she has been fighting to prevent environmentally caused cancer both inside and outside the government and scientific establishments.

As these and other stories unfold, I have noted and tried to track a number of similarities or themes. Chief among them is what Rachel Carson called a "sense of wonder." Among her dying wishes was to have completed a "wonder book" for children. She had planned to expand her children's story, "Help Your Child to Wonder" from *Woman's Home Companion*. It featured photographs of little boys, including her grandnephew, Roger, rambling through nature scenes. She had done this at her seaside home in Southport, Maine, day and night with Roger just as her mother had done with her in Springdale—following the advice in Anna Comstock's *Handbook of Nature Study* for gaining the interest of children. The text of Carson's story talks about the importance of wonder and developing a love and feel for nature as through the wide, delighted eyes of a child. Her "wonder book" was finally published posthumously in 1965 as *The Sense of Wonder*. My own 1998 edition, with photos by Nick Kelsh, still provokes gasps, smiles, happy chuckles, and even delighted talking aloud from both children and adults who read it. As Carson put it, "Once the emotions have been aroused—a sense of the beautiful, the excitement of the new and unknown, a feeling of sympathy, pity, admiration or love—then we wish for knowledge about the object of our emotional love." Carson's feeling here reflects closely the attitude of America's first major and popular nature writer, Susan Fenimore Cooper, who wrote while passing an afternoon in the woods a century before Carson, "every object here has a deeper merit than our wonder can fathom; each has a beauty beyond our full perception. . . . [T]he discolored leaves wither away upon the lowly herbs in a blessing of fertility." Such a sense of wonder is essential to an ecological perspective in which moss or microbes—or snail darters and spotted owls—are miracles with meaning of their own.

You will not be surprised then that Rachel Carson was an admirer of one of the most notable humanitarians and scientists of the twentieth-century, Albert Schweitzer. She dedicated *Silent Spring* to him and noted a number of times that she subscribed to Schweitzer's broad, nonviolent philosophical perspective, "a reverence for life." As we shall see, Carson's admiration for Schweitzer, who won the 1952 Nobel Prize for Peace; her friendship with Marjorie Spock; her acquaintance with Barry Commoner, who was one of the first American scientists to measure and oppose the effects of atomic fallout; and many other influences turned her into a strong advocate against nuclear weapons and nuclear testing. It is this sense of wonder, of reverence, and of close scientific observation in poetic prose that marks, too, the work of Terry Tempest Williams. Her modern 1991 classic, *Refuge: An Unnatural History of Family and Place*, presents an elegiac description of her favorite Salt Lake City birding spots rendered inaccessible through flooding. It is also a metaphor for searching, struggling to come to terms with loss and grief in her family life. It is Williams's emotional connection to all living things that finally gives power to her haunting recollections of her mother's and grandmother's battles with cancer, which—we learn at the end of her book—are a result of open-air nuclear testing and the distant mushroom clouds that disrupted her childhood dreams.

Imagination is another theme that runs through *Rachel Carson and Her Sisters*. It is related to, but different from, the emotional connection, the empathy, the sheer excitement of watching and feeling connected and awed by other living things. One of the most controversial and criticized aspects of *Silent Spring*—an easy mark for serious scientists some might say—is its opening fictional essay, "A Fable for Tomorrow." In it, a once delightful American town is left desolated by pollution and, perhaps, nuclear fallout. It is a place, in the words of Keats that give the book its title, where "The sedge is withered from the lake, / And no birds sing." One has to *imagine* the effects of radiation or toxic chemicals or global warming. They are not readily visible; the formulas and computer models and statistical projections that warn of invisible or distant dangers carry no feelings of the human heart. Those who love and have been moved by *Silent Spring* often recall and speak of the power of its opening fable.

The same imaginative strength marks Carson's other works; it is what made her a wildly best-selling author well before *Silent Spring*. Her first critically acclaimed 1941 book, *Under the Sea Wind*, focuses on the life stories of individual sea creatures like "Scomber" the mackerel and "Anguilla" the eel. Here Carson reflects the writing of early female naturalists and birders, who lovingly anthropomorphized birds and animals so that we could feel for them, rather than shoot them to study their bones and feathers. Among my favorites is the 1897 volume, *Citizen Bird*, by Mabel Osgood Wright. Designed for younger readers, it introduces us to the lives and habits of birds by making them a family full of cheery, cranky, and curious "citizens." The Barn Swallow, for instance, says, "Those meddlesome House People have put two new pieces of glass in the hayloft window, and how shall I ever get in to build my nest?" There is, of course, a message and some moralism here as well. But the humor, fresh perspective, and lively narrative make it go down easily, as in this early ecological entreaty from our friendly bird citizens: "House People are apt to grow selfish and cruel, thinking they are the only people upon the earth, unless they can sometimes visit the homes of the Beasts and Birds Brotherhood and see that these can also love and suffer and work like themselves."

Grounding in deep moral, religious, or scientific principles can also fortify the spine. It adds spunk—a refusal to be intimidated or beaten down. And, at times, just the right dose of righteous anger at inaction or intolerable conditions. That seems to be the case with the women I have included. Across more than a century and a half, and with very divergent personalities and pedigrees, they all refused to accept that their dreams and deeds were illegitimate, unladylike, or irritating. Ellen Swallow Richards, who earned an M.A. from Vassar College, asked boldly in 1870 to be the first woman admitted to MIT. She was not refused but was given inferior "special status." She persisted all the way to the doctoral program, where she was rebuffed because MIT did not want its first doctorate in chemistry to go to a woman. Nevertheless, Richards kept working on projects in water quality, sanitation, and health; published *Conservation by Sanitation*; became the first female fellow at the American Association for the Advancement of Science; and, at long last, like Rachel Carson, received an honorary doctorate from Smith College.

These women persevered and fought to carry out their environmental initiatives in the face of tremendous gender bias. They worked in what today would be considered hostile environments, missing recognition, promotions, or acceptance by the male establishment along the way. In spite of this, our protagonists almost all seem to share a common characteristic. It is an advantage born of adversity, nurtured by creativity and imagination, and bolstered by a wider network of women, along with some unusually sensitive, far-sighted, and sympathetic men. Since these extraordinary women were kept from the highest rungs of academic and scientific success, crossed and combined traditional disciplines, and aimed for wider and more popular audiences, they were spared the certitude, conventionality, and occasional cowardice of those crowned as keepers of the faith. Theo Colborn, for example, was unable to pursue a doctorate in zoology until her late fifties. Working for nonprofit environmental groups and supported by environmental foundations, in her sixties Colborn looked with fresh eyes at thousands of animal studies that showed the adverse reproductive and health effects of tiny exposures to organochlorines. She asked and followed up on a fairly simple question that had eluded others. Might not this mean that humans—men, women, and children—are similarly harmed? When Colborn went beyond technical writing and published her successful 1996 book *Our Stolen Future*, made appearances in the media, and advocated on Capitol Hill, she was attacked and ridiculed by the chemical industry, the *Wall Street Journal*, and other conservative defenders of the status quo. Nonetheless, her discoveries and popularization of the dangers of endocrine disruption have become mainstream environmental science.

I have arranged chapters chronologically so that in chapter 1 you can observe the slow but steady growth of the nature and conservation movement, especially the popular fascination with birds and attempts to protect them that was one of the pillars of the late nineteenth- and early twentieth-century movement that heavily influenced Carson. Chapter 2 traces the rise of the other main branch of American environmentalism—the concern with the effects of the environment and pollutants on human health that Carson was trained in and that gave such relevance and power to her work. Chapter 3 focuses on Rachel Carson and *Silent Spring* with an emphasis on the women who fought dangerous pesticides alongside her and gave her valuable support and scientific information. Chapter 4 focuses on Carson's less-well-known views as an antinuclear advocate and animal rights supporter and then segues into the work of Terry Tempest Williams, a trained naturalist and writer, whose family has been decimated by atomic fallout and who has also stretched the boundaries of ecology to include human rights and the genocide in Rwanda. Chapter 5 introduces Sandra Steingraber, a survivor of bladder cancer who, like Carson, is a trained scientist and poetic writer who has personalized and popularized the connections between environmental exposure and cancer. Chapter 6 follows Devra Davis and her friends and colleagues who have battled, through both widely acclaimed books and determined advocacy, to shift American perception and policy toward understanding cancer as an environmental disease. Her family has also suffered from deadly air pollution in her home town of Donora. Chapter 7 introduces Theo Colborn, who helped discover and popularize the threat to human reproductive health caused by even tiny exposures to endocrine-disrupting chemicals. Now in her eighties, Colborn is still fighting the dangers to human health and the environment from the growing practice of hydraulic fracturing, known popularly as fracking.

Many of the women in *Rachel Carson and Her Sisters* started careers late or had them delayed and then worked well into their eighties and nineties. Alice Hamilton, born during Reconstruction, lived to be 101 and died shortly after Earth Day in 1970. Such longevity and perseverance is an inspiration for all of us, women and men, who hope to reach and to inspire future generations. I only wish that Carson had not died of breast cancer at the frighteningly youthful age of fifty-six. I often think of what more she might have accomplished. But Rachel Carson, and the rest of the sisterhood she represents, has, nevertheless, deeply touched my life. I hope to do the same for you. It is why I write. Indeed. Writing is an act of faith.

1 ✷ HAVE YOU SEEN THE ROBINS?

Rachel Carson's Mother and the Tradition of Women Naturalists

AMONG RACHEL CARSON'S EARLIEST childhood writings is a small, handmade book, its pages pasted together and laboriously illustrated in crayon and colored pencil. It is a present to her "Papa," Robert W. Carson, done, according to her biographer, Linda Lear, with a little bit of obvious help from the strongest single influence in Rachel's life, her mother, Maria McLean Carson. In addition to the title page drawing of an elephant, Rachel's first "book" identifies her woodland friends—mouse, frog, bunny, and owl. This sense of friendship, especially with birds, is central to a slightly later childhood story in cursive script, "The Little Brown House." It is the story of two wrens, Mr. Wren and his mate, Jenny, looking for a place to nest. Rachel even draws four wren houses in the corners of the opening page. By the end of the tale, they find a dear little brown house with a green roof about which Mr. Wren declares, "Now that is just what we need."[1]

Just as these few remaining pieces of juvenilia allow us to glimpse early indications of the mature writer and scientist yet to emerge, they also offer an entrance into the world that preceded and produced Rachel Carson. We think of Rachel Carson as the mother of modern environmentalism, yet her work is as close to the nineteenth century as it is to the twenty-first. Rachel was born on May 27, 1907, the youngest of three children. Her sister and brother, Marian and Robert W. Jr., were already in the fifth and first grades. "Papa" was often absent, trying to sell insurance. So it was Rachel's mother, Maria, forced to resign her teaching job when she married, who poured all of her considerable talent and energy into her youngest, brightest child.

As soon as she could toddle, Rachel and her mother rambled most days through fields, orchards, woods, and the nearby hills of their farm in Springdale, Pennsylvania. They talked about what they saw and especially enjoyed watching the birds. Maria McLean was well-educated for her time, having attended the excellent Washington Female Seminary, near Pittsburgh. She graduated in 1887 with honors in Latin, studied a range of classical subjects that included science, and took advanced courses at nearby Washington College. Maria also passed on values from her own independent and educated mother, Rachel McLean, the studious daughter of a Presbyterian minister, who became a widow when her children were teenagers.[2]

Rachel as a Child of Nineteenth-Century Nature Writers and Naturalists

It was Maria McLean Carson who first instilled in Rachel a love of learning, an emotional identification with birds and nature, and a penchant for careful observation and scientific method. She also groomed her to become a writer. As Maria studied nature and the birds on her walks with Rachel, helped her with her first childhood booklets and stories, and encouraged her to submit her work to the children's section of St. Nicholas magazine, she was drawing on more than a half-century of work and influence arduously accumulated by a line of women naturalists, writers, educators, and conservationists who had come before. St. Nicholas magazine, whose style is reflected in Rachel's childhood writing and drawing, often featured pieces about nature by extremely popular writers like Mabel Osgood Wright and Florence Merriam Bailey. The natural science lessons and instruction Maria offered to Rachel were based on the materials that her other children, Marian and Robert, brought home from school. This widely used school curriculum was produced by Anna Botsford Comstock, a leader in the nature-study movement and author of the popular Handbook of Nature Study. It came out in 1911, when Rachel was four, and went through eleven editions between 1911 and 1936. The books and stories of Wright, Bailey, and Olive Thorne Miller, as well as the nature lessons of Anna Comstock, now mostly forgotten, are direct influences on Rachel Carson.[3]

These women, too, are part of larger networks of authors, advocates, and artists whom they read, or worked with, or knew about. As the nineteenth century progressed in America and its frontier slowly filled in and then disappeared, women began to gain better educations and the ability to travel and write, especially those in the more privileged classes. They hiked, explored, botanized, and birded, and they observed the steady encroachment of industry and urban life on the countryside. And they began to write professionally. Denied access to the heights of academia, the professions, and publishing, and restricted still by their gender, they nevertheless churned out diaries, journals, novels, children's stories and books, and magazine articles; they also published best-selling adult books, fiction and nonfiction, about nature. And, as women wrote, many also organized and advocated for the protection of wildlife, wilderness, and a vanishing way of life.

By the time of President Theodore Roosevelt, who properly gets much credit as our most-influential conservationist president, there was a rich network of women's organizations, local groups, magazines, editors, and authors who created the atmosphere in which Roosevelt was able to act to preserve wilderness and species. In these pages, we will look mainly at extraordinary women—those who achieved prominence and influence in their day yet who are now mostly forgotten. All had some scientific training or education, but they are also known as authors and advocates. Over the span of some sixty years before Rachel Carson had her first story published in *St. Nicholas* magazine, we see a shift from more descriptive writing about nature to open advocacy that something be done to save it. The combination of engaged, imaginative writing, moral values, calls to action, and writing for popular and youthful audiences led over time to a disregard for the women you are about to meet. They have been seen as sentimental, Victorian, overly moralistic, and too much given to anecdote over hard-boiled analysis and academic rigor. But their writing and lives—their influence and cultural legacy—led Rachel Carson to cherish science and narrative, imagination and wonder, the power of stories and words, and the need to take principled action.

These are values that are being rediscovered in the twentieth-first century, as hard science, technology, and a refusal to incorporate values, or consider long-term consequences, seems to have led to the continuing destruction of our natural and human environments. In some cases, like that of Susan Fenimore Cooper, we can see the evolution of a growing concern over a long career about this destruction, until, in her last years,

Cooper joined with her younger sisters in the movement to stop the slaughter of birds for fashionable women's hats. Others, like Graceanna Lewis, show the combination of wide-ranging interest in both nature and other reforms that concerned women—such as the abolition movement. Still others, like Martha Maxwell, began the development of natural history museums that presented birds and animals in natural poses and ecological settings. These are special places that Rachel Carson skipped class to visit in college, or where modern writers like Terry Tempest Williams have worked and educated new generations. Both Martha Maxwell and Graceanna Lewis were also stars at the 1876 Centennial Exhibition; they reveal how influential women were becoming conscious about asserting their new roles. Later, Mary Austin helped to develop a love of western nature and the desert that has spawned much American environmentalism, while writers like Neltje Blanchan and Mabel Wright Osborn introduced women and families in the newer East Coast suburbs at the turn of the twentieth century to a love of birds, backyard gardening, nature, and the need to preserve natural settings. By the time Rachel Carson published her first story at age eleven, she was already drawing not only on the tutelage of a loving mother, but also on two generations of women who gave birth to the first large wave of conservation in the Progressive Era. Our story begins in upstate New York, in 1850, a full century before Rachel Carson became the leading nature writer of the mid-twentieth century with her huge, instant, 1951 best seller, *The Sea around Us.*

Susan Fenimore Cooper—America's First Popular Nature Writer

Even as Henry David Thoreau labored unappreciated and obscure in Massachusetts, America's first popular nature writer was being hailed by critics and common folk alike. Susan Fenimore Cooper, the devoted daughter of famed American author James Fenimore Cooper, published *Rural Hours* in 1850.[4] It was an immediate success and preceded *Walden* by four years. We know that Thoreau himself read *Rural Hours* at least two years before the publication of his own famous book; he records in his journal of 1852, while commenting on a newspaper account of loons diving to a depth of eighty feet, that "Miss Cooper has said the same."[5]

FIGURE 1.1   Susan Fenimore Cooper in the 1850s after
the publication of *Rural Hours.* From Mary E. Phillips,
James Fenimore Cooper (New York: John Lane Co., 1913).

If mostly forgotten now, Susan Cooper was as influential in her time as
Rachel Carson in ours. *Rural Hours* went through four decades of popular
publication and revision, in the United States and overseas, where it was
published in England as *Journal of a Naturalist in the United States.* After only
six weeks of brisk American sales, the publisher, George P. Putnam, was
planning "a fine edition"; Susan's novelist father was stopped in the streets of
New York with congratulations "a dozen times a day"; and William Cullen
Bryant called *Rural Hours* "one of the sweetest books ever printed."[6] This is
important praise, because Bryant was not only the premier poet of his time
and editor of the *New York Evening Post*; he was also highly influential in
publishing and political circles. He included an essay by Susan Fenimore
Cooper, "A Dissolving View," in the groundbreaking collection *The Home*

Book of the Picturesque; or, American Scenery, Art, and Literature (1852). Her essay was placed alongside his own contribution, that of her famous father, and those of Washington Irving and other noted writers in this important illustrated literary and artistic volume that helped stimulate Americans' desire to visit, observe, and preserve America's architectural and scenic heritage. Susan Cooper's popularity grew with numerous introductions to editions of her father's work, other essays, biographical sketches, and a well-received 1854 collection of poems called *The Rhyme and Reason of Country Life; or, Selections from the Field Old and New.* With this volume that spanned the centuries and cultures, Cooper drew on her knowledge of languages along with her early sophisticated education and residence in Paris and London.

In Paris, her family lived in an elegant house at 59 Rue Saint-Dominique in the Faubourg Saint-Germain. Under high ceilings, it contained her father's fine library, a spacious salon, and a stream of accomplished American expatriates and visitors like suffragist and temperance leader Emma Willard and the artist and inventor Samuel F. B. Morse. Later known for the first telegraph, he was in Paris producing the astounding painting *The Gallery of the Louvre*, over which he labored daily at that museum; it shows in miniature most of the major masterpieces on display there. In the corner, along with James Fenimore Cooper, who saw Morse daily at the Louvre and walked home with him, we see twenty-year-old Susan (known as Sue in her youth) busily painting copies of the assembled gems. The widower Morse had been giving painting lessons to Susan, and there was always speculation about their relationship. Susan, meanwhile, had easily learned three languages. She and the entire family, as her father put it, "could prattle like natives." James Fenimore Cooper was lionized in Paris, and the family circles included General Lafayette, who introduced them to King Louis-Philippe and other wealthy and influential members of the French elite. Among Americans living in Paris, the Coopers were close to the American ambassador to France; a variety of writers, artists, and other professionals; as well as the great Prussian naturalist Alexander von Humboldt. Given her facility in languages and the world in which she was coming to adulthood, Susan Cooper was undoubtedly aware of the French naturalists who had been shaping the world's knowledge of the environment and evolution. They were congregated at the venerable Jardin des Plantes, a pleasant zoological, botanical, and archeological center, which also held lectures at its popular Musée d'Histoire Naturelle, within walking distance of the Cooper home.

But it was not so pleasant during the Parisian cholera epidemic that Susan and her family survived, though her mother did develop a bilious fever. The deadly epidemic that claimed more than eighteen thousand lives within six months began in the spring of 1832. Among the dead was the famous zoologist, paleontologist, and evolutionist Baron Georges Cuvier, whom Susan quotes in her work and may possibly have met. Given all this, it is no surprise that Cooper both extolled the fresh air and pastoral scenes of upstate New York, while showing immense erudition and familiarity with scientific and naturalist traditions.[7]

In 1887, as Maria McLean was graduating from the Washington Female Seminary, the last, shortened, and updated revised edition of *Rural Hours* was put out by Houghton Mifflin. It was followed by several essays by Cooper in 1893 that lament the decline of birds, the slaughter by the millinery trade, and the shift away from more natural small-town life. Then, in 1894, the year of her death, eighty-one-year-old Susan Cooper's final piece, a children's story, was published in *St. Nicholas* magazine. It is, of course, the very magazine that Maria Carson used to introduce Rachel to nature study and writing. Through this story, last published just three years before the birth of Rachel's older sister, Susan Cooper and her influence touches the Carson household at the dawn of the twentieth century.

Susan Fenimore Cooper was a popular, highly regarded, and influential writer whose *Rural Hours* is a clear antecedent for the fascination with rural and country life, and for Americans' wide interest in trees, birds, wildflowers, and other native flora and fauna that undergird environmentalism. Cooper heads the long line of those women whose work explains why so many Americans were upset when *Silent Spring* revealed that DDT and other pesticides were killing robins and other birds. Cooper's first published odes to robins, and others that follow hers, are the cultural legacy beneath the outrage shared by Rachel's good friend, Olga Huckins, when she wrote about watching and finding robins who had met horrible deaths from aerial pesticides sprayed into the sanctuary of her Massachusetts yard. Susan Cooper also reflects the turn of many cheerier, innocent women naturalist writers in the late nineteenth century toward early warnings of environmental destruction and the need for education, organizing, and political engagement. But *Rural Hours* and the essays of Susan Fenimore Cooper dropped out of print and out of favor.[8] Susan Cooper, like other women nature writers, had fallen victim to increased specialization in science in the academy, and to a general disregard for the romantic and Victorian values they display.

Reading *Rural Hours*

Read with fresh eyes, however, *Rural Hours* is a remarkable document—a personal journal of the progression of the seasons day by day in upstate New York with a wealth of detailed and interesting descriptions of small-town life, sadly declining Oneida Indians, and countless species of birds, flowers, trees, and precise weather observations. It is the foremother of important twentieth-century environmental writing like Joseph Wood Crutch's *Twelve Seasons* or works by Edwin Way Teale, a close friend of Rachel Carson's, such as *North with the Spring*. Born of an old family already prominent in the Federalist period, Susan Fenimore Cooper benefited from more education and access to an elite male world than most woman of her time. Through her father and grandfather, who founded Cooperstown, she straddles the worlds of colonial and Revolutionary America, when wilderness seemed to stretch forever, and the dawn of the Industrial Revolution that began to erode the old ways and old values. Although she describes, she also decries. She laments the heedless destruction of woods and forests for profit. "It is not surprising, perhaps, that a man whose chief object in life is to make money should turn his timber into bank-notes with all possible speed."[9] It is in such plainspoken moralism, linked to practical suggestions and arguments, all grounded in careful observation, that we get our first intimations of Rachel Carson a century later.

Cooper was not merely an elitist or tree-hugger, immune to the needs of a growing society. Like many of our early women writers, naturalists and scientists, she was also deeply engaged in caring for her community. At her death in 1894, she was as beloved in Cooperstown for her concern with the poor as she was for her writing. She started a free school for local, underprivileged children; was involved with "Charity House," which provided shelter for poverty-stricken families; helped establish a "Thanksgiving Hospital" after the Civil War, which is now the Thanksgiving Home for the elderly in Cooperstown; and was most engrossed by the "great work of her life, the Orphan House of the Holy Savior, familiarly known as the Orphanage."[10] It is such compassion and caring for her neighbors, growing out of her strong religious faith, that also infused Cooper's concern for rural America and its changing values, economy, and landscape. When such moral values are blended with skillful writing (she was also the author of a popular 1846 novel, *Elinor Wyllys*), her readers are moved to an emotionally felt, scientifically based concern for our civic and natural surroundings. It is an effect that modern "hard" science alone finds impossible to achieve. Consider her varied approach to life, the environment, and the economy surrounding Cooperstown.

Nature and Life in Cooperstown

Of old-growth forest pines, Cooper can wax lyrical. "And yet, in their rudest shapes, they are never harsh; as we approach them, we shall always find something of the calm of age and the sweetness of nature to soften their aspect; there is a grace in the slow waving of their limbs in the higher air, which never fails; there is a mysterious melody in their breezy murmurs; there is an emerald light in their beautiful verdure, which lies in unfading wreaths, fresh and clear, about the heads of those old trees."[11] She can also highlight important botanical characteristics. "The long brown cones are chiefly pendulous, in clusters, from the upper branches. . . . The grove upon the skirts of the village numbers, perhaps, some forty trees, varying in their girth from five or six to twelve feet; and, in height, from a hundred and twenty to a hundred and sixty feet."[12] Or, Susan Cooper can be angrily dismissive of those who destroy and exploit such trees. "It needs but a few short minutes to bring one of these trees to the ground; the rudest boor passing along the highway may easily do the deed; but how many years must pass ere its equal stands on the same spot!"[13] She goes on to say that her town might even pass away before such old-growth trees can be replaced; in fact, "no other younger wood can ever claim the same connection as this, with a state of things now passed away forever; they cannot have that wild, stern character of the aged forest pines."[14]

But, like Rachel Carson and others who followed her, Susan Cooper was a practical, politically shrewd, worldly woman who made pragmatic economic arguments as well; they foreshadow today's cost-effectiveness arguments for the value of "ecosystem services." She writes, "It has been calculated that 60,000 acres of pine woods are cut every year in our own State alone. But unaccountable as it may appear, few American farmers are aware of the full value and importance of wood. They seem to forget the relative value of the forests."[15]

After an economic argument for the dollar value of forests and their products, Cooper adds their worth to "civilization," their aesthetic and shade value, their general improvement of the quality of life. Again, she is eminently practical. "How easy it would be to improve most of the farms in the country by a little attention to the woods and trees, improving their appearance, and adding to their market value at the same time! Thinning woods and not blasting them; clearing only such ground as is marked for immediate tillage, preserving the wood on the hill-tops and rough side hills; encouraging a coppice on this or that knoll; permitting bushes and young trees to grow at will along the brooks and water-courses."[16]

Susan Cooper was primarily interested in botany, but, like Carson and others, she was what we would now call an avid birder or amateur ornithologist. The early male naturalists and bird experts explored the wilderness, shot their specimens, and were often part of military explorations. The first popular bird artists, early ornithologists—and sometimes rivals—were John James Audubon and Alexander Wilson. Obviously conversant with the wide range of naturalists, Cooper cites the work of Alexander Wilson a number of times. But she goes further, with detailed original observations of her own, like the loons noted approvingly by Thoreau. Drawing on the literature of Europe, as well as the United States, Cooper dispels myths about birds and explains the differences between the skylarks and nightingales of Romantic fame and American birds. She retells a fable of La Fontaine but adds approvingly, "In this part of the world Lafontaine would have been compelled to choose some more humble bird to teach us so cleverly the useful lesson of self-dependence."[17] She clearly wishes he might have made "acquaintance with the meadow-lark, the grass-bird, the bobolink, or even the modest little song-sparrow."[18] She can also link classical training with modern observations, as when she dispels the old belief that swallows overwintered by lying torpid in caves or tree hollows or even by surviving in the mud beneath ponds and rivers. "It is amusing to look back to the discussions of the naturalists during the last century upon the migration of swallows."[19] She refers to Linnaeus, Cuvier, and other historic naturalists with ease, even as she skewers their theories. She is a careful observer of bird behavior, even as she admires them, as in this passage on hummingbirds. "They are extremely fond of the Missouri currant . . . they are fond of the lilacs also, but do not care much for the syringe; to the columbine they are partial, to the bee larkspur also, [along] with the wild bergamot or Oswego tea, the speckled jewels, scarlet trumpet flower, red-clover, honeysuckle, and the lychnis tribe."[20]

Why Rachel Carson and Americans Love Robins

As if in anticipation of *Silent Spring*, it is Susan Cooper who first wonderfully tells of the joy of all Cooperstown at the arrival of the heralds of spring—the robins. *Silent Spring* is often best remembered for its opening fable in which a town has fallen silent and the robins have disappeared. It was the agonies of dying robins that sparked the anger of Carson's friend Olga Huckins and prompted her action. But it is in the much-read, much-loved *Rural Hours* that we first find in print the long-standing American love of *Turdus migratorious*—the American Robin.

> This morning, to the joy of the whole community, the arrival of the robins is proclaimed. It is one of the great events of the year, for us, the return of the robins; we have been on the watch for them these ten days, as they generally come between the fifteenth and twenty-first of the month, and now most persons that you meet, old and young, great and small, have something to say about them. No sooner is one of the first-comers seen by some member of a family, than the fact is proclaimed through the house; children run in to tell their parents, "The robins have come!" Grandfathers and grandmothers put on their spectacles and step to the windows to look at the robins; and you hear neighbors gravely inquiring of each other: "Have you seen the robins?"—"Have you heard the robins?" There is no other bird whose return is so generally noticed, and for several days their movements are watched with no little interest, as they run about the ground, or perch on leafless trees.[21]

Susan Cooper's cheerful description is mirrored closely by Rachel Carson in *Silent Spring* over a century later as she recounts field reports of the disappearance of robins because of pesticides. "To millions of Americans, the season's first robin means that the grip of winter is broken. Its coming is an event reported in newspapers and told eagerly at the breakfast table. And as the number of migrants grows and the first mists of green appear in the woodlands, thousands of people listen for the first dawn chorus of the robins throbbing in the early morning light."[22]

Later in her career, Susan Cooper focused even more attention on robins with a series of articles that mark her emergence as what historian Madelyn

Holmes calls "a literary ornithologist."[23] These appeared in *Appleton's Journal* in 1878 and were called "Otsego Leaves I, II, and III." They carefully describe the yearly nesting of robin families under her eaves, as well as decry the growing slaughter of birds for the millinery trade. In the nearly three decades since *Rural Hours* had appeared, Susan Fenimore Cooper had become a distinguished public figure. Here she speaks out against the growing destruction of environments and birds that are only hinted at in her first work. By 1893, she wrote "A Lament for the Birds" in the *New Harper's Weekly*, one of the most popular and respected magazines of its time. She notes the decline of various species, including the passenger pigeon, soon to fall extinct. Over the years, as the environment declined, the need for women writers to become more engaged in organizing and public advocacy grew. Providing a link to our time, Susan Cooper's last published piece, a children's story in *St. Nicholas* magazine, appeared in 1894. It is quite likely that, given her interest in the nature-study movement and her subscription to *St. Nicholas* before Rachel was born, Maria Carson was well aware of Susan Cooper and her final warnings.[24] By the time of *Silent Spring*, we can feel for ourselves the stark, painful comparison between the fairly innocent, happy writing of the young Susan Cooper and the joy of Cooperstown residents over returning robins versus "A Fable for Tomorrow," Rachel Carson's dark, prophetic warning that towns like Cooperstown—and the joy each spring that has greeted our cheeriest, most loved bird—may simply be no more.

Graceanna Lewis—America's First Female Professional and Popular Ornithologist

Susan Cooper lived and wrote in rural upstate New York and was, in effect, one of the earliest practitioners of fieldwork, the observation of actual bird behavior and other species, and their interaction with humans and a changing environment. But her younger, more urban contemporary from Pennsylvania, Graceanna Lewis, was the first woman to penetrate the male world of official ornithology and the careful, scientific study of collected specimens in a scientific institute. She was also the first to publish a recognized work of avian history, the initial volume of a planned multivolume work, called the *Natural History of Birds*.[25] But Lewis also showed the nineteenth-century female naturalist's penchant for broad, interdisciplinary study and observation. She painted, wrote poetry, was an expert on trees, and became an early proponent of nature study. Unable to get a professorship or truly scientific job, Lewis became a teacher and then a popular lecturer. Her adopted daughter, Ellen, whom the single Lewis had raised since she was a nine-year-old orphan, had by then married and become wealthy. Grateful to her mother, Ellen sent regular funds and shared her summer house at the seashore in Longport, New Jersey, with her. The two women combed and collected along the beach and were mainstays in the local branch of the Agassiz Association for Nature Study, named after the great American naturalist, Louis Agassiz, one of Lewis's mentors, and a founder and early proponent of nature study. In an eerie foreshadowing of Rachel Carson and her great ocean trilogy, Lewis, the woodland birder, was also drawn to the mysteries of the ocean and its edges. Late in life, she spent her time examining and expounding on marvelous creatures from the sea. She drew on the work of Agassiz, two of whose volumes in his *Contributions to the Natural History of the United States* were on jellyfish. She also was undoubtedly familiar with the popular and readable 1865 book *Sea Side Studies in Natural History*, drawn from his studies and written by his wife, Elizabeth Cabot Agassiz, a writer of popular natural histories, and later first president of Radcliffe College. Lewis made and published a detailed study of marine radiates, or the kind of jellyfish that washed out of the Atlantic waves and onto the Longport sands. After making careful observations of these strange, iridescent creatures, she created colorful paintings of them and identified a number of species from as far away as Florida. In 1909, she published the results, "At Longport, New Jersey, in September," for the Delaware County Institute of Science.[26]

Like Susan Cooper, Lewis, a Quaker, came from a good family with strong religious and social values. But her father died when she was three. Her mother, Esther, was quite resourceful and taught her daughter at home until enrolling her in the Kimberton Boarding School for Girls. Esther Lewis herself had a strong interest in the natural world and kept a diary that tracked eclipses of the sun, an eclipse of Mars by the moon, and comets, meteors, and auroras. She also maintained careful records of the weather and plant blooming, and discovered iron ore while examining the soil in her garden. This find led to the enterprising widow's receiving substantial sums of up to $1,700 a year from a Phoenixville company that mined on her property, paying fifty cents a ton. As with Maria Carson and her daughter Rachel, Esther Lewis was also a former teacher who then focused her talent and energy on homeschooling her daughter Graceanna. Graceanna became a teacher at her uncle Bartholomew Fussell's boarding school for girls in York, Pennsylvania, led walks to identify plants, and discovered two previously unknown wildflowers. Given early schooling by her mother, she was also able to guide her pupils in the paths of planets among the stars and constellations. But uncle Fussell's school survived for only two years; Lewis was soon living again at home, where she became friends with Mary Townsend, the sister of the noted ornithologist John Kirk Townsend. Mary had already anonymously published an 1844 book, *Life in the Insect World.* When Mary died in her thirties, Lewis, in her grief, vowed to write a book like that of her close friend, based on observations of the birds who frequented her home and its surroundings.[27]

But Lewis was also active in abolition and the Underground Railroad, and, upon the death of her mother and grandmother, became responsible for the family homestead and the orphaned child whom she adopted. It would be nearly fifteen years before she could turn her attention to the birds and her dream of emulating Mary Townsend's book on insects. She obtained Alexander Wilson's *American Ornithology,* Thomas Nuttall's *Manual of Ornithology of the United States and Canada,* and *Birds of North America* by Spencer Fullerton Baird, John Cassin, and George N. Lawrence. When Cassin, a fellow Quaker, became curator of birds at the Philadelphia Academy of Natural Sciences, Lewis was introduced to him; she fortunately found a kindly mentor. Cassin, in turn, encouraged her to introduce herself by letter to Baird, who was the assistant secretary of the Smithsonian Institution; he, too, urged her on and provided a precious microscope. Lewis then moved close to the Philadelphia Academy, published articles, discovered a new species, and finally produced the first volume of her *Natural History of Birds.*[28]

Lewis and the Uses of Adversity

We can only regret that, because of her diminished access to the Philadelphia Academy, Graceanna Lewis never was able to complete her planned multivolume magnum opus. John Cassin's mentoring of his brilliant female fellow Quaker was quite unusual. When he died in 1869, Lewis was never again as welcome in the male world of the Academy. In 1870–1871, she returned to teaching at a Friends' school. She did, however, lecture more and more widely on birds and other naturalist subjects, while remaining active in social causes, including the growing clamor for women's rights. Thus, her first real national prominence came when she spoke at the Third Congress of Women held in October 1875 in Syracuse, New York. Lewis sought to inspire other women and spoke highly of the scientific illustrations of her friend Mary Peart, whose lithographs, produced for Bowen and Company, were supplied to the Smithsonian and other institutions. Lewis praised her "marvelous accuracy in microscopic drawing, as well as her womanly desire to render her art of service to a sister scientist."[29] Lewis received support from other women as well. Her Vassar teacher, friend, and mentor, the famed astronomer Maria Mitchell, enthusiastically recommended her for a post at the college. But despite her superb qualifications, Lewis was turned down, as she noted straightforwardly in a letter to her sister, "because of the preference for a masculine representative in that chair."[30]

Blocked from an academic post, Lewis turned fully to popularizing science through lectures and reached wide, national audiences. She was a major draw for the Women's Pavilion at the 1876 Centennial Exhibition in Philadelphia, where her chart and wax model illustrating the relationship between animal species won an award, and earned praise from the judges for her "interesting and important original work." She also displayed at the 1885 New Orleans Exposition and, by the 1893 Chicago Colombian Exposition, won a bronze medal for her collection of fifty watercolors of Pennsylvania tree leaves. She then incorporated prints of them into instructional materials for schools. Active in ornithology, the study of algae, dendrology, scientific painting, and publishing in journals and magazines, Graceanna Lewis remained active despite never receiving a permanent college, university, or government post. She was influential in the widespread dissemination of nature studies and as a social reformer interested in the advancement of women. Active in the American Association for the Advancement of Women, she pushed to have professorships open to women, encouraged young female medical students, advocated for women's suffrage, and was involved in the Woman's Christian Temperance Union, which introduced the study of physiology and hygiene in schools to provide scientific underpinning for and training about the dangers of alcohol and narcotics.[31]

Despite family obligations, rejection for professorships based on gender, and the restrictions faced by single women, Graceanna Lewis's wide-ranging nineteenth-century career—from abolition and the Underground Railroad, woman's rights, and temperance to recognized ornithology and popular science—seems decidedly relevant in our twenty-first-century world. Her strong moral concern; creative, interdisciplinary study; and popular dissemination are all essential to solving the complex global environmental, health, and justice issues of our time. But if Lewis influenced the development of nature study and the role of women in and around the Eastern establishment, her contemporary and fellow star at the 1876 Centennial, Martha Maxwell, made breakthroughs not only in natural history, but in the acceptable roles and lifestyles for women scientists in a man's world. Martha Maxwell was one of the first to develop realistic natural history displays in American museums, was a strong feminist in touch with leaders like Julia Ward Howe, and was praised for her pioneering work by Helen Hunt Jackson, the defender of the American Indian, in an article for children in none other than Maria McLean Carson and Rachel's favorite, *St. Nicholas.*[32]

Martha Maxwell Heads West and Redefines Women's Work

Like Susan Cooper, Graceanna Lewis, and later, Rachel Carson, Martha Maxwell received much of her early schooling and love of nature at home. In the 1820s, when such things were unheard of, Martha's grandmother, Abigail, took off from Connecticut with her children to start a new life, despite entreaties from her husband to stay. Her daughter, Amy, married Spencer Dartt and gave birth to Martha in 1831. But Spencer soon died of scarlet fever. A distraught and weakened Amy turned to Abigail to raise her child. Martha's grandmother took Martha frequently to the woods to learn about birds, squirrels, and wildlife in general. When Amy was finally remarried—to Josiah Dartt, a gentle and studious cousin of Spencer's—Martha was further exposed to reading and nature study. In 1842, the family, including Grandmother Abigail and three of Amy's brothers, headed to Oregon. Along the way, Abigail succumbed to malaria. The Dartt family soon stopped in Wisconsin and settled there in a rough, frontier cabin. Because Josiah Dartt traveled frequently and her mother, Amy, was always sickly, Martha became quite self-sufficient. She even learned to shoot skillfully after she saved the life of her beloved young sister, Mary, by killing a rattlesnake with her father's gun.

Martha's scholarly, but poor, stepfather hoped to provide an excellent education for his bright young stepdaughter and enrolled her in Oberlin, the new coeducational college that had opened in Ohio. Josiah could not give much financial aid, so Martha skimped on everything at school, including heat. Finally, after such a strenuous first year without help, she had to withdraw. In what seemed a godsend, Martha met and married a friend of Josiah's, a fairly prosperous and older widower, James Maxwell. He convinced her first to accompany him to be the chaperone of his two children at Lawrence College, where he also offered to pay for her to attend. When James, who was forty-two and a full twenty years older than Martha, proposed marriage, she was reluctant. But her stepfather convinced her of the advantages of the match. Martha finally married James, but after a brief period of comfortable living, his business was ruined in the Panic of 1857; their daughter, Mabel, was born shortly after. The now impoverished couple headed west to pan for gold, leaving Mabel behind with grandparents. A series of ups and downs in the West followed, including Martha's running a boarding house and James having his land claim jumped by a German taxidermist who occupied the cabin they had built. The taxidermist soon departed, leaving his stuffed specimens behind. Martha was fascinated; she had also stumbled upon her future career.

While James tried various sporadic enterprises to make money, Martha returned to Wisconsin to tend to her family; there she ended up attending the Baraboo Collegiate Institute, where her stepsisters were enrolled. Inspired by the institute's natural history teacher, E. F. Hobart, and tutored in taxidermy by a local man, Martha soon developed her own innovative and more lifelike taxidermy techniques. Perhaps having learned from her grandmother's experience, Martha then set off on her own with daughter Mabel and purchased a three-acre tract in a new town colony being formed in Vineland, New Jersey. She was able to put Mabel in school while growing and selling vegetables to support herself and her daughter. This daring move seems to have stirred husband James, who finally returned as he had long promised. The small family eventually trekked back to Colorado, where Martha went camping alone or with her daughter and accompanied James, now in the lumbering business, on his trips. Throughout, she followed her newfound passion for discovering, shooting, studying, and mounting the wildlife of the frontier West.[33]

Martha Maxwell and Realistic Nature Museums

A strict vegetarian, Maxwell killed animals only in order to collect specimens; her house became a haven for wildlife, including a pet fawn and porcupine. She began to earn some money by displaying her collection and, finally, so as to earn more, opened the Rocky Mountain Museum on the main street in Boulder; it featured several rooms of native species displayed in their natural habitat. Word of Maxwell's amazing collections began to spread and reached Spencer Fullerton Baird at the Smithsonian, the same enterprising assistant secretary who had discovered and assisted Graceanna Lewis.

Maxwell began to send bird specimens to the national museum and corresponded with Baird's colleague, Robert Ridgway, one of the most prominent ornithologists of his day. She was even able to provide three specimens of the black-footed ferret that male experts had deemed nonexistent—despite being described by John James Audubon. But husband James was still no financial help. Supported by Martha, daughter Mabel, like her mother before her, went off to Oberlin College. Martha Maxwell, meanwhile, persevered, but struggled financially. In 1875, she turned for help to Julia Ward Howe, one of the leaders of the women's suffrage movement and editor of the influential *Woman's Journal*. Howe suggested showing the collection in cities and wrote an important editorial of support, saying, "her museum deserves a place in a large city."[34] Maxwell then expanded and moved her natural history museum to downtown Denver and was excited when the Colorado legislature asked her to prepare an exhibit of her collection for the Colorado pavilion at the 1876 Centennial in Philadelphia. Some attendees, including women, found her rugged garb and poses with a hunting rifle embarrassing. A vegetarian, she answered those critics of her shooting specimens (as men had always done), but who made no complaint over the death of animals for food, by saying, "I never take life for such carnivorous purposes! I only shorten the period of consciousness that I may give their forms a perpetual memory; and I leave it to you, which is the more cruel? To kill to eat? Or to kill to immortalize?"[35] Despite her stretching of traditional gender roles, Martha Maxwell achieved real fame, even becoming a kind of pop icon of the day. She was featured in numerous photographs, including a three-dimensional photo card for the

FIGURE 1.2　Martha Maxwell with her gun and palette. Courtesy of History Colorado (Scan #10038356).

FIGURE 1.3   Martha Maxwell at her 1876 Centennial display. Courtesy of History Colorado (Scan #10025478).

stereopticon, a popular, wooden, adjustable viewing device similar to the plastic ViewMaster of the 1950s, or, perhaps, in our day, a YouTube feature or nature app on today's iPhones.[36]

As Maxwell sought wider visibility for her exhibitions and work, including hopes of displaying in Paris, the financial uncertainties and strained relationship with husband James dissuaded her from returning to Colorado. Instead, in 1878 she went to Boston for further study in science, which, given her earlier peripatetic and financially strapped life, she had been unable to pursue. She enrolled in the newly opened Woman's Laboratory at the Massachusetts Institute of Technology (MIT). The laboratory was run by Ellen Swallow Richards, whom we will meet in chapter 2. The "first lady of science," Richards was the first female student and faculty member at MIT. She began innovative studies of water quality and pollution that led to Massachusetts water quality and health regulations, introduced the concept of ecology to the United States, and is one of Rachel Carson's foremothers in environmental science and public health.[37] Maxwell was thrilled at such an opportunity, excelled in her work, and met influential members of the New England Woman's Club, headed by Julia Ward Howe, and presented her work to them. When her biography, *On the Plains,* came out to positive reviews in 1879, Martha Maxwell was at the height of woman's achievements in interdisciplinary natural history. But finances remained a problem; *On the Plains* did not sell well.[38]

Despite her fame, her trailblazing gender role, and her contributions to natural history museums—especially the identification and popular public displays of bird and animal species in their ecological surroundings, Martha Maxwell was soon forgotten. Colorado had failed to pay, as promised, her costs associated with the Centennial and her collection. She was forced to make her living by displaying it wherever she could. She ultimately settled in the growing beach town of Far Rockaway, Long Island, where she died impoverished. Her daughter, Mabel, then entrusted her mother's collection to one J. P. Haskins of Saratoga Springs, New York, who was supposed to find a buyer. Instead, he displayed the invaluable collection to make money for himself. Finally, after demanding an outlandish price from the Yale Peabody Museum for Maxwell's collection, Haskins stored it inadequately. It decayed, was finally left out in the snow in a vacant lot, and ruined. Admirers ultimately gathered a few small remains of Martha Maxwell's great work. These are now housed at the Smithsonian Institution that had helped discover and display the works of this brilliant, brave, and resourceful woman.[39]

Florence Merriam Bailey—Environmental Author and Organizer

The Centennial exhibition had been marked by a display of America's rising productive capacity and the tastes of a growing middle class. As the century drew to a close, cities and early suburbs, education, and the publication of books and magazines for a broader public continued to expand. It was in this rapidly changing world that Rachel Carson's mother grew up. During this time, women became influential, if now mostly forgotten, nature writers, popular ornithologists, and the foremothers and founders of the conservation movement. Among them were Olive Thorne Miller, Florence Merriam Bailey, Mabel Osgood Wright, Neltje Blanchan, Anna Botsford Comstock, and many others. Of these, the earliest and most influential was Florence Merriam, later married to Dr. Vernon Bailey, a noted naturalist. Her brother, Dr. C. Hart Merriam, with whom she was very close, was the leading naturalist of the day and also a friend and colleague of Theodore Roosevelt, as was Vernon Bailey. Like Susan Fenimore Cooper before her, Florence Merriam was born, in 1863, into a prominent and prosperous family in upstate New York. Her father, Clinton, built a spacious house, Homewood, near his father's country estate, Locust Grove, not far from Leyden,

New York. Clinton Merriam was in both the mercantile and brokerage business in New York City and started his own firm. But according to the family story, as Florence later recalled, her father awoke one morning in Manhattan to hear a robin singing. Perhaps influenced by Susan Cooper's popular *Rural Hours* and its celebration of robins, he so longed for nature and the serenity of country life that, at age forty, he retired to rural Homewood.

Like many educated women without strong professional career opportunities, Florence's mother, Caroline, who had attended college, turned her attention to the education of her children. Caroline introduced Florence and her brother, Hart, to astronomy from the roof of Homewood and to nature and bird study in the grounds and woods around their home. Both children were also tutored by Aunt Helen Bagg, Clinton's sister, who lived nearby. An amateur botanist, she introduced them to wildflowers and other plants from the woods and her herbarium, and even taught them taxidermy. Thus encouraged, Hart avidly shot and stuffed specimens, a mostly male practice to which Martha Maxwell was an exception. By age fourteen, Hart was a leading young ornithologist with the largest collection of skins in the country. However, his collection of stuffed specimens, contained in a special three-story building that his father had built for him at Homewood, so repelled Florence that, in contrast, she quickly developed a taste for observing live birds and animals. She carried seed on her walks to attract them and, before long, lured many birds to Homewood for quiet observation and study.[40]

Meanwhile, Clinton Merriam was elected to Congress, which necessitated traveling to and from Washington. Clinton introduced his son to none other than the Smithsonian's Spencer Fullerton Baird, who had assisted Graceanna Lewis and Martha Maxwell. At age sixteen, Hart was picked to travel with the last of the Hayden Expeditions to study the geology and geography of the Rocky Mountains. He returned with 313 bird skins and 67 nests with eggs for classification by the Smithsonian. Such success made a deep impression on eight-year-old Florence. She learned much from her accomplished older brother, who published his first book, *A Review of the Birds of Connecticut with Remarks on Their Habits, Etc.* in 1877.[41] She longed to emulate him. He encouraged her to look up birds in Elliott Coues's ornithological reference that was at Homewood and answered many of her endless questions. With a wide-ranging naturalist background, attentive home-schooling, encouragement from her mother and aunt, and tutelage from her brother, now an already established ornithologist and author, in 1882, Florence headed off to the newly opened college for young women founded by Sophia Smith in Northampton, Massachusetts.[42]

Florence Merriam—First Student Environmental Organizer

It was at Smith College that Florence Merriam showed her talent for writing, nature study, and organizing. Given her lack of formal education, but her eagerness to learn, she was enrolled at Smith as a special student, not realizing that it was a nondegree status. The new women's college did not yet have a full science major, so she became steeped in literature, religion, philosophy, and the arts. During her time at college, her brother advanced in the scientific world, becoming a member of the American Ornithological Union (AOU), and starting a new book on mammals. Meanwhile, Florence Merriam tracked bird migrations for him, corresponded about birds and nature, and, finally, in her senior year, settled on birds as the topic for her article for the college science association. She had become horrified by the growing fashion for women to wear birds on their hats. She wrote to Hart and asked for all the data he could find on the use of birds for the millinery industry, telling him that she would form her article as a "preach" on the ill effects of women's fashion on bird life. When it appeared, her essay was so well received that her professor and early mentor in writing, Miss Jordan, told her that it had encouraged more Smith women to take science; it was also excerpted in two newspapers. Miss Jordan then prophetically advised her promising young student writer to think about nature writing as a career.[43]

Thanks to her brother's intervention, Florence Merriam was appointed an associate member of the American Ornithological Union, the first woman ever admitted to this professional association. The AOU encouraged the study of birds, but it also advocated for their protection from the growing devastation of the millinery trade. In her final semester at Smith, Merriam also showed a talent for organizing. She took action after a classmate waxed lyrical over having seen a spectacular hat in New York with "thirteen blue birds on it, each with a white collar and a curious topnotch."[44] These were kingfishers, striking crested birds that make a rattling call along streams, perch with a memorable profile featuring a large bill, and dive spectacularly to snatch a fish. Rather than hector or harangue her Smith classmates for wearing bird hats, Florence Merriam made a key strategic decision. She would not make fashion the enemy—a dubious enterprise for comfortable college girls. Instead, she would encourage their love and knowledge of live birds in natural surroundings while educating about their industrial-scale slaughter.

She had heard through her brother about the formation by his friend, Dr. George Bird Grinnell, the editor of the popular *Field and Stream*, of a new society, the Audubon Society for the Protection of Birds. Merriam and her friend Fannie Hardy then organized a mass meeting at which seventy-five students and five faculty members attended, presented papers, and discussed the plight of birds. Soon after, they formed the Smith College Audubon Society, one of the first of its kind. Merriam went even further and began organizing field trips to nearby Mount Tom and other birding spots, culminating in a visit she arranged for another friend of Hart's, the famed nature writer, John Burroughs. Despite rain on the appointed day, thirty Smith students followed Burroughs around Mount Tom. Merriam wrote, "The strong influence of his personality and quiet enthusiasm gave just the inspiration that was needed. We all caught the contagion of the woods."[45] Burroughs, whose *Wake-Robin* and other writings had deeply influenced Florence Merriam, returned to Smith and then to other colleges many times.[46] Before the Internet, e-mail, and social media, the first environmental grassroots network was being born. Shortly after graduating with the class of 1886 as a special student, with a certificate rather than a degree, Merriam wrote in the class letter that she had attended the American Ornithological Union meeting. There she distributed ten thousand Audubon circulars and sent articles to "about 15 newspapers, organized 6 local Secretaryships, 3 of which are now doing active work, distributed papers in all the neighboring towns, stirred the College girls up to do renewed battle, and written letters to everyone I ever heard of. My list of Audubon members is now 72."[47]

Florence Merriam Finds a National Audience and Popularizes Bird Watching

The Audubon Society numbered twenty thousand members nationally, and Florence Merriam began to write regularly for its new magazine, *Audubon*. The editor called attention to her articles, praised her organizing work, and her stimulation of wide interest in live birds. "To all such who have opportunities for field work, the example of the Smith College

FIGURE 1.4   Florence Merriam Bailey in the 1886
Smith College yearbook. Photograph by Notman
Photographic Company, Smith College Archives,
Smith College.

Society may be followed with profit."[48] Faced with what to do after college,
Merriam, who was somewhat frail, was torn between the mostly female
field of social work, which her mother had pursued, and the writing that
Miss Jordan had encouraged her to do. Writing could be accomplished at
home, allowing her to help her family. As her biographer, Harriet Kofalk,
points out, "with the wisdom of a Solomon Florence determined to combine the two."[49] She worked at Grace Dodge's Working Girls' Clubs during
winters in New York while contributing to *Audubon* and *Auk*, the journal
of the AOU. Her health remained fragile and, after coming down with
what was probably tuberculosis, her family moved to California for more
salubrious surroundings.

Florence Merriam continued to write and teach about birds, including a
summer session in Rockford, Illinois, designed for education and healthier
surroundings for factory girls associated with Jane Addams's Hull House.
It was there, in the poorest working districts of Chicago, as we shall see
in chapter 2, that female pioneers like Florence Kelley, Julia Lathrop, and
Dr. Alice Hamilton first began to systematically study the effects of urban

and industrial environments on American workers, citizens, and children. These two related environmental concerns would not be fully integrated until Rachel Carson's day. But with Florence Merriam, we can see how her intertwined values of community engagement and concern for nature are still relevant for today's environmentalists.

Returning east in better health, Merriam gathered her bird articles from *Audubon*, added new material, and in 1889 published her first book, *Birds through an Opera Glass*.[50] It was an instant hit. At age twenty-six, she had written not only a book that stressed observing live birds in the field rather than shooting them, but the first truly popular introduction to birds. It was aimed at younger readers but also at working women like those at Grace Dodge's. "I would explain to the ladies at the outset that this little book is no real lion, and that they have nothing to fear. It is not an ornithological treatise."[51] Furthermore, she wrote, "it is above all the careworn indoor workers to whom I would bring a breath of the woods, pictures of sunlit fields, and a hint of simple, childlike gladness, the peace and comfort that is offered us every day by these blessed winged messengers of nature."[52] Her book has illustrations from Baird, Brewer, and Ridgway's *History of North American Birds*. But believing that the love of birds and nature and the desire to protect them are best learned through experience, Merriam offers helpful hints to observers; she even commends her own German binoculars from Voightlander und Sohn. Neither a true field guide nor an ornithological tome arranged in the order of genus and species, but designed to appeal to and touch its readers, *Birds through an Opera Glass* helped set off today's near obsession with bird watching, an obsession shared by Rachel Carson and her friends. And, tellingly, it opens with and praises one of the most common and widely loved of American birds—the robin.[53]

Her second book, *A-Birding on a Bronco*, was published in 1896 and was based on her stays in southern California at Twin Oaks, her uncle's ranch in northern San Diego County. A charming narrative mixing personal reflections and bird observations as she rode her white horse, Canello, into the countryside, it belies the stereotype of the fainting Victorian female. *A-Birding on a Bronco* is also notable as the first nature book to use the work of artist Louis Agassiz Fuertes, then an unknown college student at Cornell; he became the leading bird illustrator of the early twentieth century.[54] Merriam's third book about birds, *Birds of Village and Field: A Bird Book for Beginners* (1898) is truly the progenitor of modern, popular introductions and guides for general readers about birds, nature, and the environment. Her narrative retains sufficient touches of her personalized, popular style, but it authoritatively covers 212 species, far more than the 70 noted in *A-Birding on a Bronco;* it is lushly illustrated with more than twenty plates by Louis Agassiz Fuertes, Ernest Seton Thompson, and John L. Ridgway, as well as hundreds of drawings of birds, in whole or in part.[55]

The Washington Influence of Florence Merriam Bailey

Florence Merriam Bailey is an important link to the twentieth century and the world in which Rachel Carson developed her career. She was the premiere female naturalist writer and organizer of her time and well-connected through her brother to the male-dominated world of science and Washington policy. Early on, Hart Merriam had praised the young Teddy Roosevelt's book on bird life in the Adirondacks, *Summer Birds.* He hailed the Harvard sophomore's contribution in an 1877 review for the influential *Bulletin of the Nuttall Ornithology Club.* Later, in his autobiography, Roosevelt notes that in college he had aspired to become a noted out-of-doors naturalist just like Hart Merriam. Before long the two men were good friends. As he rose to be the conservationist governor of New York and then vice president, Roosevelt continued to admire and stay in touch with Merriam. When he abruptly became president after the assassination of William McKinley, Roosevelt naturally collaborated closely with his friend, who by then had long been the well-known and highly respected head of the Department of Agriculture's U.S. Biological Survey (first a bureau and later a division). Roosevelt sought Hart Merriam's advice for his first presidential State of the Union address (then called the First Annual Message), went on birding and nature walks and cycling trips around Washington with him, and telephoned and visited his home regularly—on government conservation business and to get his help with bird identifications and to use Hart's vast home collection of stuffed specimens. Ultimately, it was through Merriam's division that President Roosevelt created the first chain of bird preserves with wardens to keep out poachers—even if the Audubon Society ultimately paid for the bills because of insufficient federal funds.[56]

Because Florence Merriam was extremely close to her brother, her access to and influence with the White House is astonishing by today's standards. Historians have often overlooked her role, as official documents and records tend to feature prominent male office holders and their friends and colleagues. Nevertheless, Florence corresponded regularly with Hart, for instance, until she came to Washington in 1894. She moved in with her brother, along with his wife and two children, onto the third floor of his large, specimen-stuffed home on Sixteenth Street, not far from the White House. Florence Merriam then married her brother's good friend and favorite field naturalist, Vernon Bailey, in a Christmastime wedding in 1899. Vernon worked for Hart and collected specimens on trips to the West for his Biological Survey. The young Bailey couple first lived at The Portner, a fashionable apartment at Fifteenth and U Streets, and later built a spacious brick home at 1834 Kalorama Road above DuPont Circle; they entertained frequently, and most of the prominent naturalists and scientists of the day often made their way to these convivial gatherings. Vernon Bailey had been friends with President Roosevelt since boyhood, when he had taught him how to trap wild animals. The two had also done some fieldwork together for the Biological Survey.

Unfortunately, there are no records of Florence Merriam Bailey's conversations with her brother, President Roosevelt, or other prominent figures once she moved to Washington. We do know that President Roosevelt, John Muir, John Burroughs, Hart Merriam, Vernon Bailey, and Florence Merriam Bailey were all close. But their meetings and conversations about nature, conservation, and politics—on walks, at parties, or at events—can only be surmised. There can be no doubt, however, that Florence Merriam Bailey was a well-loved, highly regarded, and important figure in these circles. We know, for example, from the diary of the locally prominent author and birder Lucy Maynard, that she, and most probably Florence, were part of a group that attended a small 1908 gathering of the D.C. Audubon Club (later the Audubon Naturalist Society) in the White House at the invitation of President Roosevelt himself. The occasion was to watch a new movie featuring the first motion pictures ever made of wild birds. Florence Merriam Bailey had mentored Lucy Maynard and encouraged her to write her 1898 book, *Birds of Washington and Vicinity: How to Find Them*. She also contributed a substantial introduction, while Lucy cites Florence as the authority on birds and thanks her as the inspiration "for this book since its inception." In the 1909 third edition, President Roosevelt himself added his White House list of bird observations. Florence Merriam Bailey also initiated and led birding classes throughout the Washington area, became a mainstay in the new Women's National Science Club, wrote for *Bird-Lore*, the magazine of the newly reconstituted Audubon Society, and worked to end the use of birds in the millinery trade through the AOU Committee on the Protection of North American Birds.[57]

Merriam Bailey's Larger Scope

Florence Merriam Bailey had already spent time in the West since her first visit there to recover her health. But later, she and her husband traveled widely in that region, working together, but independently. Vernon was the renowned specialist in mammals, Florence in birds. Following in the steps of Martha Maxwell and other women mountaineers and naturalists in the West, she helped open the eyes of the populous East Coast and its eastern, elite establishment to nature in the West. Her *Handbook of Birds of the Western United States* complemented Frank Chapman's *Handbook of the Birds of Eastern North America*. Published in 1902, Florence Merriam Bailey's volume went through seventeen editions and four revisions, and it sold well up to its final iteration in 1935. It is a true field guide, was widely used at the time, and was praised by Chapman, whose work is often cited as the forerunner of the best-selling *Field Guide to the Birds of Eastern North America* by Rachel Carson's friend, Roger Tory Peterson. Peterson's guide, revised and reissued after World War II, turned "birding" into a widespread sport and helped build the grassroots of the modern environmental movement.[58] Bailey remained prominent throughout her life. Smith College awarded her a well-earned B.A. in 1921, and the University of New Mexico followed with an honorary doctorate in 1923. Florence Merriam Bailey then won the American Ornithological Union's prestigious Brewster Medal for her 1928 book, *The Birds of New Mexico*. It is no wonder that she also became the first women ever to be voted a full member of the prestigious AOU.[59]

Rachel Carson's mother must have been aware of Bailey's work in *St. Nicholas* magazine, such as her piece on spring migration that appeared in the May 1900 issue alongside articles by President Roosevelt and the western nature writer Mary Austin. And both Maria Carson and Rachel likely saw Bailey's articles in popular and ornithological magazines like *Bird-Lore, Nature, The Auk,* and *The Condor.* But Rachel herself would have been most likely to know about Bailey's work for the Audubon Society in Washington, DC, and her many books, especially those on western birds. When Carson wrote pamphlets on the new national wildlife refuges and bird preserves for the U.S. Fish and Wildlife Service (the successor to Hart

Merriam's Biological Survey), she thoroughly researched each site that she visited and its wildlife. When she took a western trip for firsthand research, she most probably would have consulted Bailey's popular and authoritative *Western Birds*. I have not found any record of whether Carson actually met Florence Merriam Bailey, but it seems quite possible. Carson was already an active birder when she lived in Baltimore in the late 1920s and 1930s, when Bailey was at the height of her influence and fame. When Carson later moved to the Washington area in suburban Silver Spring, she became, in 1947, a formal member of the Washington, DC, Audubon Society. Florence Merriam Bailey had founded it years before, served on the board of directors, and remained a revered member into her eighties, well into Carson's time, when she died at home on Kalorama Road in early 1948.[60]

Olive Thorne Miller—Children's Author
Turns Environmental Educator

Florence Merriam Bailey was not the only female naturalist and popularizer of birds in the late nineteenth century, nor did she operate alone. She and other women before Rachel Carson were part of a wide network of mutually supportive professionals, many of whom were personally acquainted, wrote in similar publications, and shared similar goals as writers and engaged citizens. Bailey was close friends, for example, with the older Olive Thorne Miller, writing two personal appreciations of her life and work upon Miller's death in December 1918.[61] One of the most popular writers of her day, who built a successful midlife career as a children's author after raising four children, Olive Thorne Miller turned to birds even later in life. A woman of the old school, whom Bailey always properly referred to as Mrs. Miller, she wrote *Bird-Ways* and other works. The two traveled together to the still rugged West when Florence Merriam had sought more healthful climates. Their journey brought them to Utah and Mormon country, the subject of Merriam's second book in 1894, *My Summer in a Mormon Village*.[62] It was this trip that also became the basis, also in 1894, for one of Miller's most influential bird books, *The Bird Lover in the West*.[63]

Miller soon followed this success with a popular bird book for youth called *The First Book of Birds*. It is illustrated "With Eight Colored and Twelve Plain Plates and Twenty Figures in the Text" and, in its author's words, is "intended to interest young people in the ways and habits of birds, and to stimulate them to further study. It has grown out of my experience in talking in schools."[64] Mrs. Miller tells how she has been amazed at the results once children become interested in observing living birds. One boy of seven or eight, for example, convinced all the boys in his neighborhood to give up egg collecting and killing birds, and was dubbed "the Professor" by his eager, newly found followers. In writing for children, Miller interests her young readers by describing and anthropomorphizing bird families and by freely offering both practical and moral instruction. We can learn from the birds who "seem to be the happiest creatures on earth, yet they have none of what we call the comforts of life." From the outset, she makes clear that the mere scientific study of dead birds is not enough. "Men who study dead birds can tell how they are made, how their bones are put together, and how many feathers are in the wings and tail. Of course it is well to know these things. But to see how the birds live is much more interesting than to look at dead ones." Observing how birds build their nests is "pleasant." How they care for their young and seeing them first fly is "charming." You will want to know, she says, where they go in winter when they grow up. And if you go out in the field to watch and study their ways you will "be surprised to find how much like people they act." Seeing birds as much like us, of course, has a clear purpose for Olive Thorne Miller. Anyone who has studied living birds, she concludes, "will never want to kill them. It will seem to him almost like murder."[65]

Like works by Susan Cooper and Florence Bailey, *The First Book of Birds* opens by foreshadowing Rachel Carson's *Silent Spring* with the coming of spring and our iconic American bird, the robin. "Some morning a robin will appear, standing up very straight on a fence or tree, showing his bright red breast and black cap, flirting his tail, and looking as if he were glad to be back in his old home."[66] Miller leads us through the lives of these happy, industrious, and family-oriented creatures so we appreciate how they raise their young, get their colors and songs, migrate, and even help humans by destroying harmful insect pests. Finally, only after we are amused, charmed, and instructed, and when the birds have clearly become our friends, does she introduce some simple ways to identify them, to observe their structure and behavior, and how to further study them. It is the essence of the nature study method and of the stories like Miller's, Bailey's, and others that the young Rachel Carson read in *St. Nicholas* magazine. This movement, that so influenced Rachel Carson through her mother, Maria, had several origins. But the effort to save American birds from the devastation of the millinery trade was key. In their history of bird watching, Felton Gibbons and Deborah Strom report that after the National Association of Audubon Societies launched a highly successful nationwide school campaign, Junior Audubon Societies sprang up everywhere in public schools. The Illinois Audubon Society reported in 1904 that it had a membership of 1,035 adults, but more than sixteen thousand children. *Bird-Lore*, the forerunner of the current *Audubon* magazine, published educational pamphlets and bird cards, and encouraged children to submit essays and even construct birdhouses as school projects—just like the ones drawn by Rachel in her childhood story about the wren family.[67]

Mabel Osgood Wright, Neltje Blanchan, Anna Botsford Comstock, and the Rise of Suburban Nature

In the forefront of the nature study movement were other impressive women like Mabel Osgood Wright, Neltje Blanchan, and Anna Botsford Comstock. Wright was born in 1859 in New York City, the daughter of an intellectual Episcopalian minister with a wealthy congregation; he traveled in the circle of William Cullen Bryant, who, as we have seen, established the idea of the picturesque and mentored and promoted the nature writing of Susan Fenimore Cooper. Mabel Wright published more than thirty books, but those on nature and bird watching were the most memorable. When she moved to a large home in Fairfield, Connecticut, she founded the Connecticut Audubon Society, became a director of the national association, and served as the managing editor of *Bird-Lore*, where she had her own column. She turned her home into one of the nation's first bird and nature sanctuaries open to the public. It can still be visited today.[68]

In 1895, Wright published *Birdcraft: A Field Book of Two Hundred Song Game, and Water Birds*, which contained eighty plates by Louis Agassiz Fuertes. Her book went through nine printings until 1936.[69] She also wrote the charming *Citizen Bird*, coauthored with the noted male ornithologist Elliott Coues. Designed for younger readers of perhaps middle school age, it introduces us to the lives and habits of birds by making them a family full of cheery, cranky, and curious "citizens." The Barn Swallow, for example, says, "Those meddlesome House People have put two new pieces of glass in the hayloft window, and how shall I ever get in to build my nest?"[70] There is, of course, a message and some moralism here as well. But the humor, fresh perspective, and engaging narrative make it go down easily, as in this early ecological entreaty: "House People are apt to grow selfish and cruel, thinking they are the only people upon the earth, unless they can sometimes visit the homes of the Beasts and Birds Brotherhood and see that these can also love and suffer and work like themselves."[71]

When women and their families, like Mabel Osgood Wright, began to move to the earliest suburbs, a strong interest in birds, nature, and early conservation came with them. I grew up not far from the Country Life Station on the Long Island Railroad in Garden City, Long Island, where the Country Life Press of Doubleday & Co. churned out books and magazines on nature, including many of the early female naturalists. In 1897, Neltje Blanchan, another prominent nature author, wrote *Bird Neighbors: An Introductory Acquaintance with One Hundred and Fifty Birds Commonly Found in the Gardens, Meadows, and Woods about Our Homes.* It was jointly published by Doubleday and McClure Company of New York and the Nature Study Publishing Company of Chicago. It featured an introduction by John Burroughs and fifty colored full-page plates.[72] Neltje also wrote other popular books on gardening and the home, spreading an appreciation of nature into the emerging suburbs such as Garden City on Long Island. Neltje married Frank Doubleday, who eventually became the head of Doubleday & Co., which expanded to Garden City in 1910, put out *Country Life in America*, and had numerous stores and outlets. The Doubledays had homes in both New York City and in Oyster Bay on Long Island's fashionable North Shore.[73] It is close to where Marjorie Spock, reflecting this sensibility, would later convince her wealthy and prominent neighbors to sue to prevent aerial DDT spraying—a subject, like nature study and birds, that had caught the attention of Rachel Carson.

Yet another environmentalist woman, Anna Botsford Comstock, who spread nature study throughout the schools of New York State and then the entire nation, was also from a prosperous New York family. She was a Quaker, like Graceanna Lewis, and, like Susan Cooper, from upstate. Anna started her higher education at a Methodist women's college but transferred to Cornell University, where, after two years and graduating with the class of 1878, she married her zoology instructor, John Henry Comstock. She became his assistant, taught herself wood engraving, and became his text-book illustrator. In 1895, she was appointed to the New York State Committee for the Promotion of Agriculture, which sought to end the agricultural depression and the flight of country youth to the city. Comstock was named to start a pilot program in the Westchester County schools. As the program was expanded and based at Cornell, Anna spent the next ten years teaching, lecturing, writing leaflets about nature, and developing the educational program. It was these materials brought home from school by Rachel Carson's older siblings that guided Maria Carson in teaching her daughter about the value of exploring, observing, studying, and loving nature. Comstock's 1911 book, *Handbook of Nature Study*, which came out when Rachel was four, went through twenty-three editions until 1939 and was published in eight languages. It was then revised and appeared in twenty-two printings of the revised edition until 1974. It was finally reissued with a new foreword in 1986 and is still available in paperback.

Anna Botsford Comstock became the first female full professor at Cornell and was also widely influential as editor of the *Nature Study Review,* which later became *Nature Magazine;* she also helped found *Country Life in America,* which encouraged interest in rural life.[74] Anna's philosophy of teaching and her values seem to echo throughout Rachel Carson's life. At various times Comstock sought to reach agricultural students, untrained teachers, African Americans in segregated schools, children, and the general public. The goal, as she puts it, is not merely to gather facts and dispassionately arrange them. "Nature-study cultivates the child's imagination, since there are so many wonderful and true stories that he may read with his own eyes, which affect his imagination as much as does fairy lore; at the same time nature-study cultivates in him a perception and regard for what *is* true, and the power to express it." Nature study at its best, for Comstock, cultivates a love of all nature and a desire to protect it, not just study it; it is a source of health, curiosity, spiritual development, and a method to reach beyond the drudgery and specialization of traditional pedagogy. The child learns careful observation, wonder, and, perhaps most important, the value of searching for the truth and "expression of things as they are."[75] Anna Comstock's influence was so pervasive and long-lived that it stretches from the childhood of Rachel Carson and her siblings at the start of the twentieth century until my own time. I entered Stewart Elementary School in Garden City in 1948. The revised 1939 edition of Comstock's *Handbook of Nature*

Study was still in circulation in multiple printings. It was probably used by my second-grade teacher, Mrs. Small, who taught me formally about birds and nature. Its author's autobiography, *The Comstocks of Cornell: John Henry Comstock and Anna Botsford Comstock*, appeared shortly afterward.[76]

Mary Hunter Austin and the American West

Rachel Carson's only trip to the American West was in the fall of 1947 with her friends from the Fish and Wildlife Service, the artist Kay Howe and the information specialist Shirley Briggs. They enjoyed the journey immensely and the glories and birds of the West were a revelation and delight to Carson. She also would have been aware in California of one of Carmel's more famous residents, the nature writer, Mary Hunter Austin, who was friends with Jack London, Upton Sinclair, and other writers there. Mary Austin was one of the most popular writers of the early twentieth century, an iconoclast and feminist known for her shimmering, poetic prose. Her works, including her 1903 classic, *The Land of Little Rain*, and her 1932 autobiography, *Earth Horizon*, went out of print after her death in 1934.[77] But, in 1946, when Rachel Carson began to research a series of pamphlets for the general public on the national wildlife refuges for the U.S. Fish and Wildlife Service called Conservation in Action, she read widely and did thorough research on each one, including those she would visit in the West—the National Bison Range and Red Rock Lakes in Montana, and Bear River, near Salt Lake City, Utah, a locale that is central, as we shall see, to the writing of one of Carson's heirs, Terry Tempest Williams.[78]

We cannot be sure of all that Carson read, but she would have thoroughly researched material on the West for the fourth booklet in the series Guarding Our Natural Resources. General background might have included the writings of Florence Merriam Bailey and Olive Thorne Miller on the birds and life of the West, including Mormon Utah. And, as Glenda Riley has shown in her groundbreaking *Women and Nature: Saving the "Wild West,"* a wealth of women botanists, birders, conservationists, artists, and others have written about and depicted the West from a female perspective with broad, interdisciplinary strokes. They see nature as an entire, ecological entity to be appreciated and nurtured, not exploited.[79] As Linda Lear said of Carson's methods for her first series of booklets for the U.S. government, "She approached each one from a perspective that can only be described as ecological, even though that word had little common currency in wildlife science in the late 1940s." As Carson herself put it, preserving wildlife habitat meant "the preservation of the basic resources of the earth, which man, as well as animals, must have in order to live. Wildlife, water, forests, grasslands, all are part of man's essential environment; the conservation and effective use of one is impossible except as the others also are conserved."[80]

There is a kinship between Rachel Carson and Mary Austin in their ecological approach, the popularity of their nature writing, and their vivid, evocative descriptions of areas—the ocean and the desert—that had often been seen as lifeless or foreboding. Like Rachel, Mary Hunter was a somewhat lonely child with a gift for writing. Close to her father, a literarily inclined lawyer who introduced her to the classics, and to her sister Jennie, but not her mother, Mary was bereft when both her father and Jennie died early in her childhood. Her mother simply said to a friend, "If only it could have been Mary!" No wonder Mary was considered odd. When she entered Blackburn College in Carlinville, Illinois, it was expected that she would major in English. But, instead, she majored in science, saying of her English professors, "They've never written any books."[81] Clearly Mary, who was elected class poet, felt she could write on her own. After graduation, her family left Illinois to homestead in Southern California. There she entered what turned out to be a loveless marriage that produced one daughter with severe mental disabilities, who was eventually institutionalized. Throughout these trials, Mary Austin sought refuge in the mysteries of life in the desert. She wrote frequently about the Southwest, from *The Land of Little Rain* in 1903 and throughout World War I and the 1920s, when Carson was coming of age. All in all, she wrote some thirty books, including novels with feminist themes, children's verse, essays, and short stories. At least one reviewer in the *Brooklyn Eagle* recognized the importance of Austin's nature writing, saying, "What John Muir has done for the western slopes of the Sierras, Mrs. Austin does in a more tender and intimate way for the eastern slopes." And the influential literary critic, Carl Van Doren, raved that "a new degree ought to be conferred" on Mary Austin, a "Master of the American Environment."[82]

Mary Austin was poetic and mystical about the Southwest, drawing on Native American culture, older Hispanic traditions, and her own deep identification with nature, which she believed flowed through her and all living things. She had felt this since childhood when, at five years old, she had an "ineffable moment" standing under a walnut tree as if she were in communication with God. Later, when a literary celebrity, she would explain, "It is in nature that I recognize God." Such spirituality, along with her scientific background, led her to meticulous observation of the desert landscape— expressed in emotive, luminous prose. As she wrote in her autobiography of her first encounters with the desert, she was "consumed with interest as with enchantment. . . . Its creatures had no known life except such as she could discover by unremitting vigilance of observation; its plants no names that her Middlewestern botany could supply. . . . She was spellbound not to miss any animal behavior, any bird-marking, any weather signal, any signature of tree or flower." She would sit transfixed at night among dunes in moonlight, "watching the frisking forms of field mouse and kangaroo rat, the noiseless passage of the red fox and the flitting of the elf owls at their mating."[83]

It is no surprise, then, that Mary Austin could turn even the most unappealing scenes of the "lifeless" desert into both spiritual and naturalist journeys and achieve best-seller status while doing it. Consider "The Scavengers," one of the fourteen sketches that compose The Land of Little Rain. It is about death and scavengers that include coyotes, Clark's crow, ravens, buzzards (a variety of smaller carrion-eating birds), and the large vultures, among the ugliest and most symbolically scary of birds. In Austin's hands, they are all part of the landscape of drought, part of an ecosystem offered in painterly terms. They are fascinating, rather than fearsome: "The season's end in the vast dim valley of the San Joaquin is palpitatingly hot, and the air breathes like cotton wool. Through it all the buzzards sit on the fences and low hummocks, with wings spread fanwise for air. There is no end to them, and they smell to heaven. Their heads droop, and all their communication is a rare, horrid croak." The buzzards, she writes, increase in proportion to the things they feed on. This is the third of three successive dry years that have bred them beyond belief. "The first year quail mated sparingly; the second year the wild oats matured no seeds; the third, the cattle died in their tracks with their heads toward the stopped watercourses. And that year the scavengers were as black as the plague all across the mesa and up the treeless, tumbled hills."[84]

Clearly, scavengers are part of what we now call an ecosystem. Like Carson's ocean, it must be observed closely and imagined to grasp the fullness of its life and connectedness. Austin invites the reader to enter into the lives of otherwise alien beings through the kind of mild anthropomorphism used by Carson or Mabel Osgood Wright:

> Probably we will never fully credit the interdependence of wild creatures and their cognizance of the affairs of their own kind. When the five coyotes that range the Tejon from Pasteria to Tunawai planned a relay race to bring down an antelope strayed from the band, beside[s] myself to watch, an eagle swung down from Mt. Pinos, buzzards materialized out of invisible ether, and hawks came trooping like small boys to a street fight. Rabbits sat up in the chaparral and cocked their ears, feeling themselves quite safe for once as the hunt swung near them. Nothing happens in the deep wood that the blue jays are not all agog to tell.[85]

There is a message in such writing, but it is carefully cloaked in nature writing that neither berates the reader nor bores with bloodless science. Austin writes of campers who complain that wildlife or evidence of it is hard to find or see. They don't realize that things are cleaned by Clark's crow and other scavengers. "Man is a great blunderer going about in the woods, and there is no other except the bear makes so much noise." Only a very stupid animal, or very bold one, she says, cannot safely hide. Predators and scavengers are part of a system. "The cunningest hunter is hunted in turn, and what he leaves of his kill is meat for some other. That is the economy of nature, but with it all there is not sufficient account taken of the works of man. There is no scavenger that eats tin cans, and no wild thing leaves a like disfigurement on the forest floor."[86]

With the commercial success of *The Land of Little Rain* came some wealth; Mary Austin was able to finally leave her shiftless husband and place her daughter in institutional care. She moved to Carmel and eventually lived in California and in New York, where she traveled and was well known in artistic and literary circles. Austin was also active in the suffrage and birth control movements, and gave frequent lecture tours. She became deeply interested in Indian and Spanish colonial culture, producing a poem play in New York called *The Arrow Maker*. After 1924, Austin settled in Santa Fe, New Mexico, where writers and artists often visited her, including Willa Cather, who wrote *Death Comes for the Archbishop* while staying there.[87]

But Mary Austin did more than simply write about conservation; she fought for it. The Owens Valley, the site of much of *The Land of Little Water*, was soon depleted and destroyed, through political manipulation and deception, to provide water for the growing city of Los Angeles. "Mary did what she could," Austin wrote about herself in a short autobiography, "And that was too little. . . . She was stricken; she was completely shaken out of her place." A decade later, she fought against the building of the Boulder (Hoover) Dam, saying it was a "debacle," based on the greed of its backers.[88] Then, more than two decades later, she would fight valiantly again, but in another losing effort, to prevent the diversion of water from the Colorado River to insatiable Los Angeles.[89] Austin herself recognized that much of her reputation would finally rest on her nature writing about the West. She mentioned *The Land of Little Rain*, *The Land of Journey's Ending*, and *The Flock*. But she did not claim sufficient credit for herself. It would take others, like Rachel Carson's friend, editor, and early biographer, Paul Brooks, to note that *The Flock*, Mary Austin's 1906 account of sheepherding in arid country, was so full of information that it caused President Theodore Roosevelt himself to send an expert from the U.S. Forestry Service out to California to get her advice about regulations for grazing on national forest lands.[90]

Women's Wider Influence

Each of these women and their friends and colleagues were part of a larger social movement as well. They were not content to simply study nature, write about it, and inspire youth. They were part of the emerging Audubon Society and among its founders and leaders. When the Milliners Association finally offered some restrictions on the trade in bird feathers to Audubon and the American Ornithological Union in 1900, male leaders like Frank Chapman and Joel A. Allen of the AOU were inclined to accept, even though the agreement would have only restricted North American feathers and would have required Audubon and the AOU to refrain from opposing the trade in edible and game birds and in domestic fowl. It was Mabel Osgood Wright who offered the most stinging critique in her column in *Bird-Lore*. She pointed out that the compromise allowed the continuation of some of the very practices that the Audubon Society had been formed to end. She concluded sarcastically, saying of the proffered deal, "We will not interfere with you even if you cover your hats with birds so long as they are labeled 'killed in Europe!' "[91]

By the early twentieth century, then, a broad, decades-old, nationwide conservation movement was already in place—one that is much richer than the traditional narrative of a few early male leaders and writers like John Muir or Gifford Pinchot influencing Teddy Roosevelt, our first conservationist president. And within that movement, broadly defined, women played a substantial role. Like Muir, Burroughs, or Pinchot, our popular and influential women authors are merely the most visible peaks of a wide range of female conservation efforts and consciousness. As Glenda Riley puts it, quoting the 1890s humorist Josh Billings, "Wimmin is everywhere." A little digging, she writes, shows that, "The vast numbers of environmentally minded women in the late nineteenth and early twentieth century are almost unbelievable. . . . [T]heir stories trickle forth from any research or archival collection. With a bit more effort, the trickle turns into a flood. From birders, botanizers, and other naturalists to travelers and tourists, women were present in force. Almost always, women's interest in nature turned into concern for the environment's future."[92]

It should be noted, too, that Gifford Pinchot, who had claimed credit for his own conservation ideas, also praised the widespread efforts of women in the Progressive Era in his 1910 book, *The Fight for Conservation*. These included campaigns by the Daughters of the American Revolution, the Pennsylvania Forestry Association, Minnesota women who preserved national forests, and California women who, after a nine-year fight, set aside the Calaveras Big Trees. As environmental historian Carolyn Merchant points out, women had developed organizations in the nineteenth century that paved the way for the conservation effort. Women's clubs, along with the women's rights and abolition movements all exposed women "to the political process and the public arena." Particularly important was the General Federation of Women's Clubs, founded in 1890, which had been active in forestry since the beginning of the century with cleanup, beautification campaigns, and the purchase of lands for conservation. It had some 800,000 members nationwide who were kept abreast of Roosevelt's and Pinchot's conservation policies and state affiliates that took on specific campaigns. Notable among these was the preservation of old-growth forests in California. The California Club had been founded in 1897 at the home of Mrs. Lowell White of San Francisco, after a failed suffrage campaign. In 1900, she and the club took up the cause of forestry and merged with women's clubs throughout the state to form the California Federation of Women's Clubs. Mrs. Robert Burdette became the first president and noted in her impassioned opening speech, "While the women of New Jersey are saving the Palisades of the Hudson from utter destruction by men to whose greedy souls Mount Sinai is only a stone quarry, and the women of Colorado are saving the cliff dwellings and pueblo ruins of their state from vandal destruction, the word comes to the women of California that men whose souls are gang-saws are meditating the turning of our world-famous Sequoias into planks and fencing worth so many dollars."[93]

By 1908, the General Federation of Women's Clubs was central to the conservation effort passing resolutions, lobbying, and taking active roles in National Conservation Congresses held from 1909 to 1912. There was a Women's National Rivers and Harbors Congress formed in 1908 that soon had twenty thousand members. In 1909, the DAR, under the leadership of President General Mrs. Matthew T. Scott, who had defeated the conservative wing of the organization, brought seventy-seven thousand members to the conservation fight and supported conservation committees in every state. The head of this DAR campaign was none other than the mother of Gifford Pinchot. The DAR battled for the preservation of Appalachian watersheds, the Palisades, and Niagara Falls, which was threatened by the water usage of electric companies.[94]

As the Great Depression and World War II overshadowed conservation efforts, the pioneering writing and work of women on behalf of the natural environment was mostly forgotten. But their contributions created a cultural and political legacy that newer generations of women could build on. American attitudes about birds, backyards, and the benefits and beauty of some form of wilderness had already been shaped. It should be no surprise, then, that after World War II and the widespread use of pesticides like DDT, Rachel Carson and other women were once again at the heart of the struggle to save birds, restrict chemicals, and redefine American's views of technology, progress, and even nature itself.

2 ✢ DON'T HARM THE PEOPLE

Ellen Swallow Richards, Dr. Alice Hamilton, and Their Heirs Take On Polluting Industries

IT IS 1958. Olga Owens Huckins has turned her home in Duxbury, Massachusetts, into a bird sanctuary. She is following in the footsteps of Mabel Osgood Wright, nearly three-quarters of a century earlier, whose popular book, *Birdcraft*, and creation of a home bird sanctuary in Fairfield, Connecticut, had started a trend. But when Huckins finds robins, writhing and dead, in her beloved oasis, she is outraged. Ever since tales of a town celebrating the return of the robins in spring had swept the nation in 1850 with Susan Fenimore Cooper's best-selling nature book, *Rural Hours*, the robin has been the most loved American bird.[1]

But now something in American life has gone horribly wrong. A woman of action, Olga Huckins writes an angry letter to the *Boston Herald* to refute claims by a Massachusetts bureaucrat that no harm is being done from the aerial spraying of the miracle chemical of World War II—DDT. She sends a copy of her letter to her good friend, the writer Rachel Carson, in Washington. Huckins describes the robins in her yard: "We picked up three dead bodies the next morning right by the door. They were birds that had lived close to us, trusted us, and built their nests in our trees year after year. . . . On the following day one robin dropped suddenly from a branch in our woods. We were too heartsick to hunt for other corpses. All of these birds died horribly, and in the same way. Their bills were gaping open, and their splayed claws were drawn up to their breasts in agony."[2]

When Rachel Carson receives Olga's letter and *Herald* clippings, she, too, is heartsick and outraged; she, too, takes action. Thanks to her mother and women like Cooper, Wright, Florence Merriam Bailey, and Anna Botsford Comstock, she has always loved birds, especially robins, since childhood. Like these and other women, Carson is also part of an organized movement. As a member of the DC Audubon Society, she knows what to do; she has already been part of efforts to protect birds by saving critical habitat. She calls the congressional offices of the House and Senate committees that are investigating wildlife damage.[3] She initiates a series of steps that will lead to *Silent Spring*, still best remembered for its haunting opening fable, a dystopian version of Susan Cooper's lovely descriptions of Cooperstown and its joyous spring greeting of the returning robins. In Carson's version, there are no robins, wrens, or other birds. Just eerie silence. Ultimately, the robins, as well as larger, iconic birds like the American bald eagle, the pelican, and the peregrine falcon will be saved because of her writing and that of the women who came before. But Rachel Carson and *Silent Spring* are so effective not simply because a love of robins, birds, and nature has been aroused by a wondrous writer. She makes a compelling emotional and scientific case for why an American town could turn into a nightmare—without birds *or* people. Rachel Carson is the inheritor of two traditions in American history in which women have played prominent roles. The first, as chapter 1 describes, is a movement that has developed the love of birds, nature, and the writing of natural history; it has sought to protect and conserve America's wildlife and wild places. The second tradition, as we will discover in this chapter, is a movement driven by a concern for the health of children and communities and the effects of chemicals and other pollutants on our lives.

Shortly after receiving the exchange of letters in the *Boston Herald*, Carson begins intense study of the dangers of all pesticides. She calls up the U.S. Department of Agriculture (USDA) and its assistant administrator for the Plant Pest Control Division in the Agricultural Research Service (ARS), W. L. Popham. He is in charge of the government's fire ant eradication program. She asks Popham for a number of ARS research bulletins on pesticides and for the percentages of dieldrin and heptachlor, among the newest and most deadly toxins being used by the USDA, in aerial spraying to destroy fire ants.[4] Rachel Carson is not merely operating out of emotion—her imagination and ability to feel and portray the agony of robins—though there is that. She is also a trained scientist. Like other women before her,

FIGURE 2.1   Ellen Swallow Richards, "First Lady of American Science." Courtesy of the MIT Museum.

most of them little known and without much credit, Rachel Carson is about to launch a research investigation into the consequences for human life—families, children, citizens, not just birds—of environmental contamination. Fortunately for us, she combines serious training in environmental science and public health along with her renowned ability to write. She is the first major figure in our time to truly combine these talents, to weave together two separate strands of environmentalism that had long been separated. In this, she most resembles and draws on a tradition that started with another foremother from Victorian times, the "First Lady of American Science," Ellen Swallow Richards.

Ellen Richards, along with her somewhat younger contemporary, Dr. Alice Hamilton, epitomizes those women in the late nineteenth and early twentieth centuries whose interest in science and social concerns, along with an ability to write and feel keenly the awful effects of the growing industrial age in America, led them to focus their attention on the plight of poor workers, neighborhoods, and urban landscapes where the more obvious victims of pollution were Americans themselves. Trained in chemistry and medicine, despite the obstacles for women at the time, Richards and Hamilton, and their friends and colleagues, like Harriet Hardy and Mary Amdur, who came after them, laid the groundwork, with their science, writing, and advocacy, on which Rachel Carson built her case against the dangers of chemical pollution. They are her foremothers in focusing on the effects of pollutants on human health that is the less remembered, though equally important warning at the heart of *Silent Spring*. These women also issued calls to action, were part of broader movements, and were attacked for their views and their gender. But their achievements are all the more remarkable as a result. And like Carson's environmental sisters in chapter 1, their integrated views of ecology and social justice, and their denial of admission to the highest levels of the male establishment allowed them to question and see environmental issues in entirely fresh ways.

Ellen Swallow Richards—Mother of Ecology and the First Lady of American Science

Ellen Swallow, who later married Robert Richards, a mineralogist at MIT, first introduced Ernst Haeckel's German concept of *oekology*, or ecology, to the United States and was so proficient in so many disciplines and practical organizing that her hand is behind a stunning series of developments. Her biographer, Robert Clarke, pointed out in the early 1970s how she helped to launch associations, to found disciplines, and to pioneer health and environmental studies. But Richards still lacks direct credit or even recognition for many of her achievements. Among them is her initiation in 1881, and rescue from closure in 1888, of the Woods Hole Marine Laboratory. This is where, in the 1930s, Rachel Carson furthered her postgraduate education and got her firsthand knowledge of the ocean. There is also Ellen Richard's training and mentoring of the early male leaders of the American Public Health Association and the Johns Hopkins Schools of Medicine and Public Health.[5] These are where the young Rachel Carson found female colleagues and some supportive men as she pursued graduate studies and taught summer sessions in biology at Johns Hopkins and the University of Maryland.[6]

Ellen Swallow's Early Years and Education

Like Rachel Carson and many nineteenth-century women environmentalists we have looked at, Ellen Swallow Richards first developed her love of nature, of learning, and of science at home. She was born, in 1842, to Peter and Fanny Taylor Swallow on a farm in Dunstable, Massachusetts, not far from the New Hampshire border and the White Mountains. Her parents were both well-educated for the time. They had met as teachers at the Ipswich Academy in New Hampshire. But Fanny was frequently ill and Peter turned to farming. He had time for only part-time teaching. Ellen seems to have inherited her mother's frailer constitution. Believing that Ellen's health would be improved by fresh air and outdoor activity, and critical of the low quality of the schools available in Dunstable, the Swallows homeschooled their precocious daughter until she was sixteen. She roamed the nearby hills and woods and streams, learning a love of nature and its healthful benefits. From her father, she learned history and logic; from her mother, numbers and letters. Later, she learned mathematics and literature from both.

In 1859, recognizing the gifted Ellen's growing need for more education, the Swallows moved a few miles away to Westford. There Peter opened a general store. At age seventeen, Ellen was enrolled at the Westford Academy—her first formal education. She excelled in languages—Latin, French, and, the rarely taught German. But her genius for mathematics was also recognized; she was asked to tutor other students in math as well as in Latin. Meanwhile, she increasingly helped her father at his store, becoming adept at business, keeping books, managing inventory, and traveling to Boston alone to buy and barter for supplies from wholesalers who commented on her quickness. Her mind clearly moved at a torrid pace. As a child, she was taught to speak more slowly in order to be understood as her thoughts raced along and words tumbled forth. At her father's store, she observed the buying habits of customers, read books constantly, and when coming and going between school and store made copious notes, drawings, and maps of nature—noting topography, stream flows, the growth and condition of plants, and animal behavior.

When Ellen graduated in 1863, the Swallows moved to Littleton, where Peter opened a larger store, with a post office attached. His daughter also worked there. Then, at twenty-two, she took her first teaching job, but had to give it up to care for her mother, who was ailing again. When Fanny was fully recovered, Ellen Swallow struck out on her own, moving to the city of

Worcester. Here she sought more schooling but also worked with a Deacon Haywood with whom she visited jails and insane asylums, and also taught a class of boys at a mission school. But, once more, in 1866, she was called home to help. Having experienced freedom, teaching, and service to society on her own, the brilliant young Ellen Swallow, for the first and last time in her life, soon suffered from depression. Later, she wrote to a woman undergoing similar circumstances, "I lived for more than two years in purgatory." Of her feelings without meaningful work or further education, Ellen Swallow wrote simply, "Tired, tired, tired . . ."[7]

The clouds lifted in 1868 when Ellen Swallow heard about and was admitted—given her experience at age twenty-six—as a third-year student at Vassar College. It had opened just a few years before in Poughkeepsie, New York. Swallow had the good fortune to be mentored by one of the truly great pioneering American women of science—the astronomer Maria Mitchell. Mitchell recognized and encouraged her new student's talent and drove her to be a meticulous scientist. Swallow wrote home, "I think Father would be delighted to see Miss Mitchell lecturing me. . . . I ignored one one-hundredth of a second in an astronomical calculation. 'While you are doing it you might as well do it to a nicety' she told me." Soon the young Ellen Swallow identified star clusters and meteors that even Mitchell had missed. Her fellow students noted her immense powers of concentration, her desire for more and more knowledge, her reading while walking. Again she tutored in math, Latin, and German, and brought rocks, fossils, soil, and animal and plant specimens home to study under a microscope. She entered into a new world that would become her domain—the analysis of water and its contents invisible to the naked eye. Admired and mentored by both Mitchell and by her male chemistry professor, A. C. Farrar, Swallow ultimately chose to focus on chemistry. She explained that it was "a leaning toward social service" that pulled her toward the more immediately practical field of chemistry.[8] At Vassar, Ellen Swallow was noted for brilliance in several fields. She sharpened her political and organizing savvy; took up surveying, dressed in a daring athletic bloomer suit; and helped on a project of early meteorological recordings for the Smithsonian Institution.

When Swallow graduated from Vassar in 1870, she was mature, independent, and ready for intellectual adventure. But challenging jobs were hard to find, especially for women, so she again worked for her father. By now he had moved the family to Worcester and opened a new venture in artificial stone. Ellen Swallow helped set up the business and examined the composition of the new materials, products of growing post–Civil War technology. But still craving broader horizons and challenges, she signed a contract to teach astronomy in Argentina and raised the funds (the first of many serious fund-raising efforts at which she excelled) to get herself there. But as she was packed and ready to go, war broke out in Argentina; her job was cancelled. "Hired and fired before I started," she wrote. She corresponded widely to find jobs in chemistry shops or pharmacies, ultimately applying to Merrick & Gray, a chemical company in Boston. Its answer changed history. We do not hire women, was the reply. But the firm was kind enough to suggest she try the new Institute of Technology in Boston.[9]

Ellen Swallow—First Woman at MIT

When Harvard had rejected all requests to incorporate technological training and applied science into its classical curriculum, the Massachusetts Institute for Technology (MIT) opened in 1865. By 1866, MIT was deeded some land on Boylston Street, moved to new facilities there, and opened its first chemistry laboratory. Swallow was advised by friends, however, that MIT did not admit women. Undaunted, she sent a letter seeking admission, with Vassar's Maria Mitchell and A. C. Farrar as references. When she did not hear back, she applied for a job with Booth & Garrett, a Quaker chemical firm in Philadelphia. They, too, turned her down, but wrote that they hoped to open new fields to women. They offered to take her on as a student if she would pay $500 for the privilege. It was not possible for her.

All roads to her goal of being a chemist seemed closed when, on December 3, 1870, a letter finally arrived announcing that MIT had received her request; it had been discussed for more than four weeks, and no decision had been reached. It would be considered again at the next meeting which, as it turned out, was just a week later. The faculty met on December 10 and decided to admit her as a "special student in chemistry." Her presence at MIT was described as "only to be considered an experiment." Then, on December 14, Ellen Swallow, as the first woman ever admitted to MIT, or a higher science institute of any kind, received an official, pleasant letter from the president of MIT, J. D. Runkle. He was eager to meet her, was impressed by support from Mitchell and Farrar, and noted kindly that "you will have any and all advantages which the Institute has to offer without charges of any kind." She was thrilled. Only later did she learn that the university was hedging its bets and that her special status meant she was not formally admitted. She was on trial. The first woman would have to be spectacular—and on her guard. Later, she noted that if she had known the true circumstances of her admission and status, she would have refused.[10]

It was not long before Ellen Swallow impressed some of the doubters at MIT. President Runkle quickly caught the determination in her eye when they first met. He escorted her to the gender-segregated area where she would work as assistant to the only other woman at the school, a Mrs. E. A. Stinson, a staff member who was "Assistant in Charge" of the chemical supply room. Runkle urged her to take good care of the new special student. After Swallow had left, Mrs. Stinson commented that the young woman looked rather small and frail for such a demanding course. "Yes," President Runkle replied, "but did you see her eyes? She will not fail."

By her second semester, however, after interruptions from illness and another return home to tend to her dying father (whose leg was amputated without anesthesia), Ellen Swallow was about to be asked by the faculty to leave MIT. It is unclear how she was able to stay on, but after being released from her women-only segregated status, she was able to impress her chemistry professors. Soon she was working alongside William Ripley Nichols, a noted chemist, who had left Harvard for MIT. Nichols had been asked by the Massachusetts Department of Health to assess the water quality of Mystic Pond. He had also been part of the group that had opposed the admission of women to MIT. But now, in 1872, Nichols chose his best student, Ellen Swallow, for the difficult task of analyzing samples from a second, larger water quality study. She wrote, "A new task has been put into my hands by a professor who does not believe in women's education. . . . I have made about 100 analyses so far and that is only part of my daily duties. I . . . prepare my lessons . . . evenings." Swallow became expert at the difficult work required and even expanded her knowledge by acquiring water samples from around the world. To his credit, when the report was concluded in 1873 (the year she was awarded a master's degree by Vassar for a breakthrough study of vanadium), Nichols gave due recognition to his brilliant female student. "Most of the analytical work has been performed by Miss Ellen Swallow. . . . I take pleasure in acknowledging my indebtedness to her."[11]

Already an expert in the analysis of trace contaminants in water; working alongside and translating critical German texts for a young mineralogist, MIT's rising star, Robert Richards; and with her complete competence in

FIGURE 2.2   Ellen Swallow Richards sampling water quality in Massachusetts. Courtesy of the MIT Museum.

several fields recognized by a previously skeptical faculty—Ellen Swallow took the obvious next step. She applied for MIT's brand-new PhD program in chemistry. The record is revealing. Swallow was turned down and happened to see papers marked "A.O.M." She asked a sympathetic chemistry professor, John Ordway, what it meant. "Actium Omnium Magistra," he replied, meaning a unanimous decision by higher-ups. Always bright and outspoken, she saw through the bureaucracy's cover, "An Old Maid, I see it." It was a bitter disappointment to her. Years later in his autobiography, her husband, Professor Emeritus Robert Richards, recalled that Ellen had wanted a doctor's degree more than anything else. "She was treated for some time as a dangerous person. . . . It seems to me . . . that some of the difficulties may have arisen from the fact that the heads of the department did not want a woman to receive first D.S. in Chemistry."[12] After time had elapsed, an officer of the MIT Corporation was completely candid about "the tradition of a long controversy with Miss Swallow, President Runkle and a few faculty members on one side, and the rest and all the Corporation on the other. Why [they] asked, 'should a female take up scientific studies reserved for men? Woman's place is in the home . . . and if one woman is admitted, others will follow; and think of the disastrous effect on the young men.' "[13]

Ellen Swallow and Women's Science Education

Following such a discriminatory dismissal, Swallow's determination to succeed, the initiatives she undertook at MIT despite continuing low status and the denial of a doctoral degree, and the new directions she carved outside of academia are more than admirable; they are astonishing. Always interested in opening doors for other women, Swallow, along with her friend Bea Capen, a science teacher at the new Girls High School in Boston, were members of the Women's Education Association (WEA). Together, they sought to train more female teachers in science. They were keenly aware of the need and the serious lack of opportunity for young women to study chemistry, when, in 1872, a young woman came to Boston to prepare for a career in medicine. She could not be admitted to medical school without chemistry. But such training was hard for a woman to find. One of Ellen Swallow's professors, James Crafts, took the woman's case to Dr. Samuel P. Eliot, Capen's employer at Girls High. He then proposed to the WEA that "[i]f the Association will pay for the materials and instruction," she could study chemistry in the evening at Girls High. The WEA was interested. In February 1873, not one, but sixteen young women began a class in advanced chemistry under Ellen Swallow and Bea Capen at Girls High.

Not content to rest there, Swallow began to lobby for coeducation and the training of women in science at MIT itself. As she wrote to Robert Richards, "After all, I'm not the only one. There are a great many able young women coming through college. They will be very important to science education if we allow them to have one first."[14] She followed up such entreaties, along with some flattery for Professor Richards, by making him acquainted with promising female students. She began to bring groups of them to MIT to watch the handsome young professor carefully construct hand-blown glass chemistry apparatus. Richards was not in favor of coeducation at this point and was clearly wrestling with the concept that Swallow had been so skillfully lobbying and arguing for. In his diary, he reviewed the arguments for and against coeducation that she had put before him. His opposition slowly faded, even as his interest in Miss Swallow increased.

In June 1875, Ellen Swallow married Robert Richards, the scion of a well-connected Boston family, professor of mining engineering, and head of MIT's new metallurgical laboratory. Shortly afterward, having laid the groundwork over several years after rejection from the MIT doctoral program, she approached the WEA for funds, writing that "I have reason to believe that if you will provide the funds, [they] will provide space for a woman's laboratory at the Institute."[15]

Ellen Swallow Richards then raised an additional $2,000 needed to set up the lab at MIT. With funds assured, the government of the Institute voted to admit special students in chemistry "without regard to sex." But, again, the school was cautious. "Special student" status applied only to Richards's chemistry students, not to general admission. The chemistry lab itself was gender segregated and constructed in an old building that badly needed renovation. A member of the Corporation recalled later that "the cautious and reluctant authorities placed her and a small band of disciples in a sort of contagious ward located in what we students used to call 'the dump.'" But ever a pragmatist, Richards was happy at having finally opened the gates of science to women and MIT to coeducation. In November 1876, twenty-three women, mostly teachers, began scientific training in chemistry. The first women's science laboratory in the world had opened. Professor John Ordway was technically put in charge. But its daily operations were run by the woman who had made it happen—Ellen Swallow Richards.[16]

She then began to train an entire generation of promising young women who came through her laboratory and classes. By 1878, Ellen Richards was made an assistant lecturer without pay, becoming MIT's first female faculty, even if without official, full status. She assisted young women with finances, sought scholarship funds for them, helped find housing and even part-time jobs for them. She also instituted new disciplines and courses in science and generally served, without title, as dean of women. Though underappreciated by MIT, in 1878 Ellen Richards was promoted from member to full fellow in the Association for the Advancement of Science (AAAS), an honor usually reserved for outstanding, doctoral-level men. Throughout this time, she worked on and studied various scientific disciplines in addition to her specialty in water quality. She studied and discovered minerals, geology, evolution, and more. In 1879, having pioneered the field of engineering, she was admitted as the first woman and only female member of the American Institute of Mining and Mineralogical Engineers.[17]

Ellen Richards Discovers Ecology and Changes Public Policy

Given her wide interests and abilities, her concern for the education of women for broader social purposes, and her need to serve many functions as the "dean of women" without formal status at MIT, it is not surprising that Ellen Richards crossed disciplines, took new innovative approaches to research and pedagogy, and was drawn to emerging concepts in Europe. So it was that as a young wife, she and her teaching friend Bea Capen set off with Professor Robert Richards on his research trip to Europe in June 1876. Robert was going to visit mines and smelters. Bea and Ellen sought out universities and their laboratories. On this trip Ellen visited the University of Jena and the laboratory of Ernst Haeckel, the man who had proposed evolutionary theories before Darwin and had introduced the new concept of *oekologie* in 1873. Jena was also a source of fine glass and optical equipment that interested Robert as well. Haeckel seems not to have been in Jena when Ellen visited, but her laboratory tour was conducted by Carl Zeiss, an enthusiastic and devoted friend and follower of Haeckel and his theories. It was Zeiss who provided Haeckel with his instruments and who founded a firm that was among the first to produce fine, lightweight prism binoculars used by American naturalists, male and female, in the nineteenth century and to this day. Zeiss was so impressed by Ellen Richards that he provided instruments for the Women's Laboratory, which then introduced courses in biology taught by herself and Professor John Ordway.[18]

With her interdisciplinary, ecological perspective, the social and economic aspects of food and nutrition next led Ellen Richards to become, in effect, the founder of the American consumer, nutrition, health, and right-to-know movements. As she put it in an 1879 speech: "Perhaps the day will come when an association will be formed in each large city or town with one of their number a chemist. Some similar arrangement would be far more effective . . . than a dozen acts passed by Congress. The power of knowledge is appreciated by manufacturers. They take advantage of every new step of science. The woman must know something of chemistry in self-defense. If the dealer knows his articles are subjected to even simple tests, he will be far more careful to offer the best."[19]

For Ellen Swallow Richards, knowledge for women meant not just greater access to higher education, but building associations and networks to disseminate and interpret information to consumers, homemakers, and the public. In a burst of organizing, she founded, within a very short period, an amazing set of nonprofit organizations that still play a role today. In 1876, she organized the Boston chapter of Vassar Alumnae; she then went national in 1881, helping to found the Association of Collegiate Alumnae— now the American Association of University Women (AAUW)—where, declining to run for president, she chaired the Executive Committee.[20] Among her efforts there, Richards soon arranged for a scientific survey on the health and psychology of college-educated women to contest the widely held belief at the time that college education literally hurt women's health. The results proved otherwise and shifted the entire framework and terms of debate over what women could do. Massachusetts officially published the findings and declared that: "it is sufficient to say that the female graduates of our colleges and universities do not seem to show, as the result of their college studies and duties, any marked difference in general health from the average health likely to be reported by an equal number of women."[21]

Richards continued to promote women's full potential in industrial training, in social and mental hygiene within a broad context, and in early child hood education. She expanded the Society to Encourage Study at Home with the nation's first college extension courses, including the sciences and home laboratories where students received microscopes and other instruments. They studied zoology, botany, physical geography, math, archeology, chemistry, and the forerunner of environmental health called sanitary science. Students young and old were involved all across the nation and often had direct contact with her. One urban student wrote, "Now every little twig has meaning to me." Another, who became a leader in public health, wrote, "just what I wanted and needed . . . I only hope someday to know enough to help some girls as you are helping me."[22]

In addition to all this, Ellen Richards traveled widely, visiting and lecturing at schools and colleges, writing up her ideas and lessons for broader audiences in *Health*, her first of many books. It was aimed at women with depression and advocated books, poetry, and mental stimulation, as well as fresh air, pure water, and good food. Then, the studies she had commissioned and led through the American Collegiate Alumnae were compiled and published by her friend Marion Talbot as *Home Sanitation*. Finally, in 1882, Richards published her first book for women in the general public, *The Chemistry of Cooking and Cleaning*. It remained popular and in print for over a quarter century and filled a gap between educated women and then isolated housewives. Its goal was to educate and empower women and build a network of informed, engaged citizens who could demand a decent environment—natural, physical, and social—for themselves, their communities, and children.[23]

Again with the backing of the WEA, Richards went to the Massachusetts State Board of Health, Lunacy, and Charity and proposed a scientific examination of food. Leading a band of colleagues, she went to stores in more than forty cities, from rich to poor neighborhoods, buying goods, interviewing grocers, and bringing food samples back to the lab to study. The object was to find and expose unhealthy adulteration. Richards found misleading statements and adulteration from manufacturers, unclear labeling, and widespread consumer ignorance of what they were actually buying. She and her lab persevered in this consumerist and health approach, finding cinnamon tins filled with mahogany sawdust and colored mustard thinned with cheaper substances like starch. She published all this in popular journals like the *New England Farmer* until, between 1882 and 1884, with particular note of the work of Ellen Swallow Richards, the Massachusetts legislature passed the nation's first pure food laws.[24]

Ellen Richards, "Sanitary Chemistry," Water Quality, and Woods Hole

Such a record of accomplishment was still not sufficient, however, for MIT. By 1883, new specialized fields were emerging within biology, thanks to the work of Louis Pasteur, Robert Koch, and others. The germ theory of disease was born, and the causes of tuberculosis, diphtheria, typhoid, and cholera were identified. MIT needed to beef up its credentials, especially in bacteriology. Meanwhile, Richards's old supporters were departing, and a new president, Francis Amasa Walker, had replaced the man who took a chance on her—J. D. Runkle. Walker reorganized MIT, picked William Sedgwick, a young graduate of Johns Hopkins, to teach biology, passing over both John Ordway and Ellen Richards, who had built the program. The women's laboratory was to be torn down and a new coeducational one built. Richards was very happy to hear this news until she discovered that she would not be working in the new lab (which first reverted to training primarily men). Her unpaid job and other duties were eliminated, and Ordway soon resigned. At age forty-one, the first lady of science was totally unemployed. "I do not know that I shall have anything to do or anywhere to work," she wrote; "everything seems to fall flat and I have a sense of impending fate which is paralyzing."[25]

Somehow, Ellen Richards was not deterred. During 1883, she helped the young Sedgwick redesign courses and build the biology program; she introduced him to the ideas of the great European labs with which she was familiar; and she shared her expert knowledge of air, water, and sanitation. The neophyte Hopkins man called her his "great teacher." Then, her fortunes turned once again. Her friend E. O. Atkinson, president of Boston Mutual Manufacturers Fire Insurance, made Richards his chief industrial chemist, the first woman ever to hold such a position. Then MIT gave William Sedgwick permission to open a bacteriological program and to open a new Sanitary Chemistry Laboratory that would focus on sanitary science and public health. Nichols was nominally the head, but Ellen Richards was put in charge and officially added to the MIT faculty as an instructor in Sanitary Chemistry.[26]

In 1886, when Robert Richards developed typhoid on a trip to the West, William Ripley Nichols died while in Europe. Again, Ellen Richards was passed over to officially head the Sanitary Chemistry lab. Dr. Thomas Messinger Drown was appointed chief, and she joked, "still the number two man." But, once more, she made herself indispensable and buried any hard feelings. And so it was that Ellen Richards undertook and completed the most comprehensive water quality survey in the United States. She and the late Nichols had already planned for it when funds were allotted by the Massachusetts legislature. For more than two years, with a team under her control, Richards surveyed the water supply and sewerage of some 83 percent of the state population, personally carrying out testing and analysis of forty thousand samples. She literally worked night and day, supervising, analyzing, improving laboratory techniques and procedures, and finally, tabulating the results herself on a large map of Massachusetts. She drew lines on the map—isochors—that connected and revealed areas in Massachusetts where water pollutants had increased to a similar level, as marked by higher levels of chlorine than along the coastline. Her exact and elegant mapping, soon called the Normal Chlorine Map, became an important new tool for tracing contamination and gave rise to concern about ocean pollution as well.

Given the newness of the field of water pollution, Ellen Richards sent an organizing letter to prominent naturalists seeking advice, especially because the WEA was no longer able to financially support the seaside laboratory in Annisquam that she had initiated. A group of twenty-one experts from

MIT, Harvard, Tufts, Smith, Wellesley, Williams, and the Shaw School of Botany in St. Louis gathered at the Boston Society of Natural History and voted to form a corporation to oversee the laboratory at Annisquam; it was then transferred to them by the WEA and moved to Woods Hole. Two of Ellen Richards's former students were made trustees, as was Sedgwick, "to receive funds." Richards was behind most of this, and her water survey soon led to Massachusetts's first water quality laws and sewage treatment, including the new Lawrence Experimental Station in Lowell. Biographer Robert Clarke properly complains that she should have gotten more credit for her work, which went instead to prominent men. But Richards seems to have been more interested in results than accolades. Following the Sanitary Survey and recognizing the need for ongoing monitoring and analysis of state water quality, Massachusetts opened a new State Water Laboratory. It was placed at MIT, in the cubicle of Ellen Swallow Richards.[27]

Establishment Backlash

Ellen Richards continued to teach, mentor, consult, and write more and more books. Throughout, she attempted to foster broad, interdisciplinary ideas of ecology, and later, a similar, but somewhat narrower concept focused on what we might call healthy lifestyles; she labeled it euthenics. But science continued to specialize and professionalize during the 1890s. As a new wave of economic depression and conservative reaction set in, it only became harder to connect consumerism, public health and sanitation, medicine, and environmental contamination with human health and welfare. Traditional gender roles began to reassert themselves so that chemistry, physics, and engineering became increasingly male dominated. Nature and the life sciences, along with social work and concerns related to home and community, were relegated as the preserve of women. To further complicate things, as economics, gender roles, and politics shifted around her, Ellen Swallow Richards was first seen as too radical. But then she was ultimately driven into the ever-shrinking and increasingly gender-based box of home economics—even though she herself had first conceived of it as a way to give women, as well as men, power and scientific understanding for the modern world.

Trouble began when Richards made an important, pointed, and powerful speech at the 1896 American Public Health Association (APHA) convention held in Boston that year. The APHA is a venue that (as we shall see in chapter 3) also heard from Dr. Alice Hamilton, the founder of occupational and environmental medicine; was attended in the 1930s by Rachel Carson; and is one where contemporary women like Devra Davis (the focus of chapter 6) have also presented. Ellen Richards strode to the podium and delivered a bold challenge in direct language on that day more than one hundred years ago. With the press in attendance, she charged the municipal officials of Boston, parents in the city, and "the great tax paying public" with "the murder of some 200 children per year." It is the term *murder*, of course, that is so incendiary, though something like *involuntary manslaughter* or *negligent homicide* would not have soothed things much. What she was talking about was public health statistics and deaths related to environmental conditions in Boston's public schools. She told the assembled health scientists that "[f]ully half of [Boston's] school houses [are] deleterious to health. . . . 5,053 cases . . . of disease among students, are caused by illegal conditions: no ventilation, open sewer pipes, filthy toilets, and buildings where 41 per cent of the floors went unscrubbed since laid." She went on to underscore that only 27 of 168 schools in Boston had working fire escapes. Not only children were at risk and being sickened and killed; Boston's teachers also had the nation's highest death rate. The result of this courageous exposé was mainly inaction and growing hostility. But Richards kept at it, giving more speeches, writing in national publications, and finally turning to the National Education Association (NEA) to bring even greater pressure. When Boston remained unmoved, she formed a citizens' committee to go over the heads of Boston politicians. She organized educators and opinion leaders around the state and managed in 1897 to get the Massachusetts legislature to hear their complaints. Ultimately, a number of school reforms were passed and spread nationwide. Again, years later, a member of the MIT member corporation and leading educator, James Phinney Monroe, wrote of her four-year crusade, "In . . . conducting the campaigns [she showed] in a superlative degree her special qualities: pugnacity, tempered by extraordinary tact; thoroughness, leaving nothing to chance."[28]

But a backlash had begun. The publisher of her nutrition periodical, *New England Kitchen Magazine*, withdrew his support. When Thomas Drown left MIT to become president of Lehigh University, the state water lab was removed from MIT's (and Richards's) control and incorporated into the statehouse. When William Sedgwick, whom Richards had graciously assisted and groomed, started a national society of bacteriologists, she was not invited. Her allies at Woods Hole were voted out, and as she put it, "All the enterprises in which I am involved seem . . . to be in an uncertain condition."[29]

Yet one more time, however, Ellen Richards remained undaunted and kept organizing. She began a series of conferences at Lake Placid, New York, with the help of Melvil Dewey, the inventor of the Dewey Decimal System. This network of educators and reformers became known as the Lake Placid Conference, which at a meeting in Washington, DC, in 1908, voted to become the American Home Economics Association (AHEA). By 1909, W. O. Atwater's graduate school at Wesleyan University joined, bringing in new academic disciplines, including agriculture, dairying, animal and plant physiology, landscape architecture, and other postgraduate specialties. It also attracted large state universities like Illinois, Ohio, and Iowa and prestigious private schools like Cornell and Yale. These moves increased male membership as well. The National Education Association included home economics in its annual meetings beginning in 1910, and the AHEA, through Richards, was working closely with the APHA. Together, they urged greater attention and appropriations for investigating nutrition at the Department of Agriculture. The Congress soon agreed.[30]

The result of these and other organizing, educational, and lobbying efforts by Ellen Swallow Richards was profound; she laid the groundwork for the American consumer, environmental, and public health movements that still undergird American environmentalism and from whose influence Rachel Carson emerged. The interdisciplinary and reform-oriented movement that focused on the domestic health and well-being of Americans was conceived of by Richards as part of a broader concept of ecology; it touched all levels of American society. Ellen Richards had influenced and built correspondence courses, rural schools, elementary and high schools, colleges, and graduate and professional schools. She had affected rural Grange organizations, as well as the one million members of the General Federation of Women's Clubs. Now in her late sixties, she came to the 1910 annual meeting of the AHEA that she had founded with a ten-year plan for its future— along with her resignation, to allow the growth of newer leaders and to give herself time for still more projects. The association voted her "President for Life."[31] Soon after, and not long before her relatively early death at age sixty-seven, the first lady of science, who was never allowed to complete a doctoral degree in chemistry at MIT, was awarded an honorary Doctor of Science degree from Smith College. It is almost the same scenario that would be enacted a half century later with Rachel Carson.[32]

Dr. Alice Hamilton and Occupational and Environmental Health

Rachel Carson seems not to have considered a career in medicine, but she was deeply influenced by it and by public health, or what today we might call the environmental health movement. She found women friends, colleagues, emotional support, and concern for the impact of the total environment on human health at the Johns Hopkins Schools of Medicine and Public Health, at the Woods Hole Marine Laboratory, and when she taught biology part-time at the Johns Hopkins undergraduate campus and at the University of Maryland School of Dentistry and Pharmacy. As we will see in chapter 3, by the time of *Silent Spring*, Carson was corresponding with activists, advocates, and health professionals nationwide; they were pursuing the links between the new set of persistent organic compounds and cancer and other human diseases. To trace direct links from early pioneers to Rachel Carson is difficult. There is no documentary evidence that Rachel Carson ever wrote or said, "I was influenced by and admired the work of naturalists and advocates like Florence Merriam Bailey and the activist environmental and health scientists Ellen Sparrow Richards and Dr. Alice Hamilton and their friends and colleagues." But she might as well have. Her research on pesticides and both animal and human life drew on official government reports and agencies and on people whose work, in the late 1950s, grew out these antecedents. At the center was Dr. Alice Hamilton, whose public work in establishing occupational, industrial, and environmental medicine begins around the time of Ellen Richards's death in 1911.

Alice Hamilton is still virtually unknown to most people; she is rarely mentioned in environmental histories. As the founder of occupational and environmental medicine in the United States, Hamilton falls into categories other than "the environment": medical, industrial, and labor history; Progressive Era reform movements; and social movements like peace and civil liberties, where she was quite active. Rachel Carson drew no such distinctions, especially in *Silent Spring*. Both Alice Hamilton and Carson opposed nuclear weapons and the dangers of fallout and radiation exposure; both were concerned with the effects of newly synthesized chemicals on humans.

Although Carson drew on Marion Gleason's 1957 toxicology text for *Silent Spring*, she would most likely have been familiar with some version of Alice Hamilton's pioneering 1934 work, *Industrial Toxicology*, which was revised and updated by Hamilton and her younger Harvard colleague Harriet Hardy in a 1949 second edition.[33]

Alice Hamilton was born in New York City in 1869 but raised by her prosperous, educated family in Fort Wayne, Indiana. She shares with many other women pioneers a practicing Protestant family; strong education in languages and literature; a love of nature study and science at home as well as at school; a doting, educated mother; undergraduate education at a women's college; and a Victorian sense of the need to be purposeful and socially useful. Her sisters were also accomplished. Edith Hamilton is known to many as the author of texts such as *The Greek Way* that were used in elite colleges to introduce students to classical studies. The bright, talented Hamilton sisters were sent east for prep school at Miss Porter's in Farmington, Connecticut, and then forced to decide between the few careers open to women. Edith went to Bryn Mawr College and hoped to teach there but instead became the long-tenured, successful head of the Bryn Mawr preparatory school in Baltimore. Her widely read books were all written later in life, beginning in her fifties.[34]

Alice, on the other hand, chose medical training, which was not as regulated or specialized as now. As she described her choice, "I chose it because as a doctor I could go anywhere I pleased—to far-off lands or to city slums—and be quite sure that I could be of use anywhere. I should meet all sorts of men [people], I should not be tied down to a school or college as a teacher is, or have to work under a superior, as a nurse must do."[35] She studied physics and chemistry with a high school teacher in Fort Wayne and, starting in 1890, attended "one of those third-rate medical schools which flourished in the days before the American Medical Association reformed medical teaching."[36] Then, in 1892, she convinced her father to let her attend a "real" medical course of training at the University of Michigan Medical School, where she was admitted as a special student. Alice Hamilton flourished there, with professors trained in Germany. Michigan was one of the few schools that offered biochemistry, bacteriology, pharmacology, and physiology and had more laboratory work than lectures. Women were still required to sit off to the side, but Michigan had been coeducational for some twenty years; Hamilton found little of the overt sexism she found later in eastern schools. Dr. George Dock, professor of medicine, put her on his staff in her last year, where she seized every opportunity to take time away from surgery and other clinical subjects to work in his clinical laboratory. There she used a microscope and chemical analysis of tissues to determine the presence and causes of disease in individual cases. This was the "new" medicine, pioneered by Sir William Osler, which began to explore disease in humans by observations *inside* the human body.

While in medical school, Hamilton spent summers in Mackinac, Michigan, with her sisters and had her first taste of "emancipation"—an introduction to radical ideas from Chicago friends who discussed Henry George's *Progress and Poverty* and supported the anarchists in the Haymarket bombing trial.[37] After just three semesters at Michigan, she was able in 1893 to graduate early, given her previous experience and her impressive study and work. At the suggestion of her mentor Dr. Dock, Hamilton followed the University of Michigan Medical School with two brief internships in Boston and then a year of study in Germany, which she undertook with her sister Edith. It was in Boston and her nine months at the New England Hospital for Women and Girls that she first saw big city life and poverty. "What interested me most," she wrote, "was the life down in the Pleasant Street dispensary where I worked with people of thirteen different nationalities and each new call was an adventure." She also worked alongside a Russian medical intern, Rachelle Slobodinskaya—later Dr. R. S. Yarros, one of the founders of the birth control and social hygiene movements—who had been a revolutionary, fled Russia, and worked in sweatshops in New York. Hamilton was clearly deeply affected by these experiences and later became a close friend and colleague of Dr. Yarros at Hull House in Chicago. But this connection did not develop until after her training in Germany, which included microscopic study of cholera and other bacilli and a range of experiences that gave her a long-standing interest in foreign affairs.[38] Upon returning from Europe in 1896 and finding no positions available for trained bacteriologists or pathologists, Alice Hamilton enrolled for the winter as a resident at the Johns Hopkins Medical School in Baltimore, where Rachel Carson worked not quite thirty years later.

Like Rachel Carson, Hamilton found the new medical school, especially given the prevailing attitudes of the day, to be relatively open to women. "Baltimore was not gay and colorful as were Munich and Frankfurt, but the men I worked with accepted me without amusement or contempt or wonder, and I slipped into place with a pleasant sense of belonging."[39]

FIGURE 2.3   Dr. Alice Hamilton, who graduated from
the University of Michigan Medical School in 1893.
Bentley Historical Library, University of Michigan.

Alice Hamilton Combines Medicine and Social Work

By the summer of 1897, Hamilton was finally able to find a full-time position
teaching pathology at the Women's Medical School of Northwestern University in Chicago. She was happy to have found a medical post there, but
also to "realize a dream I had had for years, of going to live in Hull-House."[40]
Between her time at Miss Porter's and the University of Michigan Medical
School, Alice and her sister Agnes had discovered the writings of the social-
ist Richard Ely. Then they read about the growing settlement movement
and finally heard Jane Addams, the founder of Hull House, at the Methodist Church in Fort Wayne. Alice Hamilton recalled, "I only know it was
then that Agnes and I definitely chose settlement life. Years later we carried
out our resolve: Agnes went to the Lighthouse in Philadelphia and I found
myself a resident of Hull-House."[41]

The influence of Hull House and the settlement movement has been
well told elsewhere. Suffice to say that as the twentieth century was about
to dawn, many of the women who preceded Rachel Carson and combined

an interest in nature, science, and politics were drawn into its orbit. Ellen Swallow Richards visited Hull House frequently while at the 1893 Chicago Exhibition to display her science and nutrition-oriented kitchen. She then introduced the study of nutrition to Hull House, which spread it among poor women in the city. Richards's friend and earliest biographer, Caroline Hunt, was in residence at Hull House for two years, 1894–1896, just before Alice Hamilton arrived and taught at the nearby Lewis Institute. Though direct evidence is lacking, it seems reasonable to assume that Hamilton, Hunt, and Richards all crossed paths. Furthermore, in 1891, Florence Merriam Bailey had already introduced poor women to bird and nature study in the countryside at the summer institute run by Hull House in Rockford, Illinois, Jane Addams's hometown. Given that Bailey worked at a settlement house under Grace Dodge during winters in New York, she probably would have been aware of developments at Hull House and other settlement projects before she contracted tuberculosis and headed west.[42]

Alice Hamilton spent some years at Hull House before focusing on occupational and environmental medicine. Slowly, she found her métier in a combination of scientific and medical observation and concern for the poor people of Chicago and the unhealthy conditions to which they were exposed. She worked alongside the so-called garbage lady, Mary McDowell, who fought to improve working-class homes; the crusading Florence Kelley, who fought for consumer protection; and the calm, compassionate Julia Lathrop, who during the New Deal, would head the first Child Labor Division and with whom she became close. Hamilton queried male colleagues at medical school about industrial exposures and disease. Most knew little or found the subject somehow suspect. To them such questions smacked of "socialism" or feminine sentimentality. Hamilton was neither impressed nor intimidated.

By contrast, in Europe, scientists, reformers, and parliaments had been dealing with occupational disease. When Hamilton read the work of John Andrews, who had just completed a study of phossy jaw in American match factories, she became more determined to get involved. She had already observed limp arms and colic from lead exposure and other diseases. Then she came upon and read the 1902 British exposé *The Dangerous Trades* by Sir Thomas Oliver, which described horrid working conditions and occupational diseases. She read extensively on the subject for several years at the Crerar Library in Chicago and learned more and more about the dangers of lead, mercury, benzene, carbon disulphide, and other toxic substances. But the science literature always talked about Europe. Then John Andrews actually came to Hull House to discuss with her his first-ever study of phossy jaw in the United States. When published in 1909, it led to major reforms in the match industry and to federal legislation ending the disfiguring disease forever. Alice Hamilton finally knew exactly what she needed to do with her life and her skills.[43]

Alice Hamilton, Environmental Health Policy, and Community Epidemiology

Alice Hamilton's break came when Charles Henderson, a professor of sociology at the University of Chicago, recommended her to Illinois Governor Charles S. Deneen to head up a study of Illinois workers. Henderson had been impressed by sickness insurance benefits given to workers in Germany. Required to give cash benefits to help laborers with their illnesses, employers quickly improved conditions in factories. This, in turn, led to fewer illnesses. Henderson's goal was to similarly impress Governor Deneen. When asked who Dr. Hamilton was, he described her this way:

> Alice Hamilton is a resident worker at Hull House, where she has lived for the past thirteen years . . . She knows the different industries and the workers in Ward Nineteen better than any other person. She speaks German, French, Italian in addition to her native English. With her gift for friendliness, she goes into the shops and homes of these immigrant workers and talks with them. For the past two years she has read everything she can find in foreign journals concerning occupational diseases in other countries. She is a scientist concerned about social problems and in my opinion the most fair-minded physician I have ever known. A survey of industrial diseases in Illinois would be the first study of its kind ever in the United States.[44]

In 1908, Governor Deneen appointed Alice Hamilton as the only woman and the chief medical investigator of a new nine-member commission. Then, beginning on March 1, 1910, she headed her own study of lead poisoning, which she had observed many times. Her familiarity with poor families, her work in the neighborhoods, and her caring approach allowed her to get information that would have otherwise been almost impossible to gain.[45] Alice Hamilton was helping to establish a key principle of what would become community-based environmental health epidemiology.[46] Without accurate medical records, addresses, reports, or other data on lead poisoning, she had to resort to leads and suggestions from those she met until she was able to track, observe, interview, and write up sufficient numbers of cases in factories and homes from Hungarian, Polish, Bohemian (Czech), and other workers exposed to lead. She uncovered 304 cases of lead poisoning from various dangerous trades utilizing lead in Illinois. In one case, she observed men applying lead-based paint to enameled bathtubs, but found no evidence, as the factory manager had assured her she would not, of lead exposure. Only when she was able to talk once more to one of the Polish workers in private did she discover that the exposure came within a different factory on the far northwest side of Chicago. It was there that finely ground enamel was poured over a red-hot metal tub and filled the air with lead-laden dust. Her reports described the effects of exposure: "A Hungarian, thirty-six years old, worked for seven years grinding lead paint. . . . [H]e had three attacks of colic, with vomiting and headache. I saw him at the hospital, a skeleton of a man, looking almost twice his age, his limbs soft and flabby, his muscles wasted. He was extremely emaciated, his color was a dirty, grayish yellow, his eyes dull and expressionless. He lay in an apathetic condition, rousing when spoken to and answering rationally but slowly, with often an appreciable delay, then sinking back into apathy."[47]

Given such unprecedented documentation, Hamilton was soon speaking at the First National Conference on Industrial Diseases, sponsored by the American Association for Labor Legislation. Then the Deneen commission sent her to the Fourth International Congress on Occupational Diseases and Accidents in Brussels, where she was warmly received. Next, she went on to the Twenty-Third Annual Meeting of the American Economics Association to speak on the need for workforce compensation laws. She was becoming a well-known, expert, and trusted figure on industrial toxic exposures. Before her Illinois work was even completed, she was invited by Charles Neill, U.S. commissioner of labor, who had heard her speak in Brussels, to carry out a national, trade-by-trade, state-by-state study of the lead industry.[48]

Alice Hamilton soon began her independent travel and investigations. She uncovered 358 cases of lead poisoning, 16 resulting in death, among 1,600 lead workers. But because some five times that many workers passed through white lead factories, the actual numbers were clearly much higher. This tweed-suited, upper-class doctor had visited each of twenty-two factories twice—once to review conditions and another to examine the workers. This all had to be done with only an identity card as a medical consultant of the Department of Labor and her considerable poise, charm, and sympathy. Working without pay until her report was finished, Hamilton traveled from Omaha through Illinois and on to the main cities of the East. She returned to Hull House to write her first of many publications that would follow—a seventy-page report published by the Bureau of Labor called *Industrial Poisons*.[49]

In the meantime, the Illinois legislature received and studied the report of the Illinois commission that Hamilton had chaired. It pointed out existing laws that needed to be followed such as protecting workers from fumes, gases, and dust. They were impressed and quickly passed a strong, new state law signed by Governor Deneen on May 26, 1911. It required manufacturers to get insurance for complaints and insurance companies to see that the companies met all legal requirements, including adequate lunchrooms and lavatories and medical care for their workers. It was the first of its kind in the nation. But the courts were not as progressive as the legislature and ruled in 1935 that occupational diseases were "a stranger to the common law." By this time, sixteen other states had included occupational diseases in their compensation laws, but only in 1936 was the Illinois legislature able to pass a Workmen's Compensation Diseases Act acceptable to its state court.[50]

Alice Hamilton continued her nationwide investigations for the Department of Labor, moving from each sector of the lead industry to the next—smelting, printing, and battery production. At each visit, perhaps because of her personal style, perhaps because she was a woman who appeared to be unthreatening and friendly, she was able to gain admission and trust. In Joplin, Missouri, where Missouri, Kansas, and Oklahoma come together, an area noted for its concentration of lead mines, mills, and smelters, she visited Smelter Hill. There, three young owners were operating a lead smelter they were quite proud of and had asked her to visit. As biographer Madeleine Grant—who was able to interview both Alice Hamilton and her sister Margaret, who lived together in their nineties until Margaret's death in 1969 at age ninety-eight—describes it, Smelter Hill was "a dreary spot: the unpainted shacks where the people lived were sagging; not a tree could be seen. Only large pyramids of crushed rock—refuse from the lead mills—met the eye. Clouds of fine sand blowing off the pyramids filled the air."[51] Grant's account also points up the importance of talking to and working with community people when investigating environmental health conditions: Hamilton stops and chats with a local woman in Joplin who has already heard all about her coming to town. "We all knew you was coming. They've been cleaning up for you something fierce. Why in the room where my husband works they tore out the ceiling, because they couldn't cover up the red lead. And a doctor came and looked at all the men and them that's got lead, forty of them, has got to keep to home the day you're there."[52] The next day, after inspecting the plant, Hamilton met with the three young managers who were waiting for her presumably favorable report. "I'm sorry," she is reported as having said, "I cannot say anything about conditions here, because I have not really seen it as it is, only as it has been staged for me personally." The men look guilty as schoolboys and she bursts out laughing. She somehow wins them over and they agree to let her see doctors' reports and make permanent reforms.[53]

Alice Hamilton also was able to use her connections with the wealthy to make reforms as well. Or perhaps we should see her as a forerunner of stockholder activism. Appalled at conditions she found at a factory making Pullman railroad cars where lead-laden dust from scraping paint filled the air and only an old Civil War–era company doctor was available, she talked to Jane Addams about it and her disgust with Pullman. Addams put her in touch with Mrs. Joseph P. Bowen, a wealthy friend and supporter of Hull House and a major Pullman stockholder. After Mrs. Bowen's formal complaint, Hamilton met with Pullman managers, and major changes— including a modern surgical department, eye specialists, and a medical department to supervise some five hundred painters—were in place before the end of 1912.[54]

Alice Hamilton Joins the Peace Movement—and Harvard

With the outbreak of war in Europe in 1914, Alice Hamilton, along with Jane Addams, became deeply engaged in the American and international peace movement. She was a forerunner of Dr. Benjamin Spock, who led antinuclear, antidraft, and Vietnam War protests during the late twentieth century. She attended peace congresses and supported peace plans put forth by Addams and the new Women's International League for Peace and Freedom. When the war's terrible destruction and carnage had ended, she and Jane Addams were asked by the Quaker humanitarian Herbert Hoover, before he became president, to inspect and report on conditions in Europe, especially in Germany where hunger was widespread. The two women friends and other colleagues stayed on for some time, providing the reports and impetus for Hoover's postwar relief efforts. The war stimulated interest in occupational diseases and, in 1919, Alice Hamilton was invited to join the faculty of the Harvard Medical School. Loath to give up her work on industrial disease or move from Hull House, she negotiated a part-time arrangement and became Harvard's first female faculty member of any sort. On March 10, 1919, Hamilton was appointed assistant professor of industrial medicine. There were protests that she was "pro-German" because of her humanitarian work and peace efforts, and one wealthy female patron withdrew financial support from the medical school. But another man, Mr. Hammar, the owner of a white lead factory in Missouri who was grateful to her because of her clean-up efforts, convinced the Lead Institute to endow a three-year study of lead at the medical school.

Nevertheless, Harvard remained cautious and conservative about this breakthrough appointment of a woman who had made the national news. Alice Hamilton, it was understood, would not be allowed to use the all-male faculty club from which faculty wives were barred; she was not to attend football games; and she was not to sit on the platform with male faculty at commencement. Despite Harvard's antiquated culture, Hamilton persevered, as Ellen Richards had done at MIT, because she found the work and the chance to set larger precedents more important. She swiftly became a respected and popular teacher, and when the department of public health became the new Harvard School of Public Health, she headed the study of industrial poisons.[55] Continuing her work with Hull House, the Department of Labor, and Harvard, Hamilton was now the leading American expert on diseases caused by exposure to industrial pollutants like lead, beryllium, carbon monoxide, carbon disulphide, benzene, and mercury. She brought the same painstaking and compassionate methods to each problem and by 1925 published the most important book in the field, *Industrial Poisons in the United States.*[56]

Alice Hamilton and the Fight against Benzene, Mercury, and Cancer

Devra Davis, whom we will meet in chapter 6, has carried on work today started by Ellen Richards, Alice Hamilton, and Rachel Carson in linking environmental exposures and human health. She admiringly describes Hamilton's work on benzene and its adverse health effects, first set forth in a series of articles in 1916 and 1917. Davis portrays what she calls an ongoing scientific game of hide-and-seek as manufacturers and their political and medical allies consistently hide, downplay, or try to deny clear evidence of harm from toxic chemicals brought forward by consumers, workers, and their reform champions. The British medical journal *Lancet* first described benzene as a "domestic poison" as early as 1862, not two decades after it was first produced on an industrial scale. Then, Alice Hamilton's work is filled with references to case reports in French, German, and other journals, as well as with early American cases of benzene poisoning reported in 1910 at Johns Hopkins. Hamilton carefully describes the effects of case after case, including a strong man in a rubber factory who dipped wooden forms into a tub of rubber in solution with benzene and staggered home, only to fall unconscious when he got there. In Uppsala, Sweden, young women working in a bicycle tire factory bled from the nose, mouth, and gums, hemorrhaged, and had uncontrolled menstrual bleeding. Four of these young Swedish women died within a month of becoming sick.

Under the auspices of the U.S. Bureau of Labor Statistics, Hamilton then visited forty-one wartime plants in 1917 where explosives were produced using benzene-based compounds. Many workers were poisoned. Some, like the Uppsala girls, bled to death from massive hemorrhages. Between 1919 and 1940, at least thirty-three publications advocating the replacement of benzene with safer chemicals were released. Many of them, according to Davis, were from the National Safety Council and Dr. Alice Hamilton.[57] Despite these efforts, by 1948, the American Standards Association, an industry experts group, still maintained that a person could be safely subjected to 100 parts per million of benzene over an eight-hour work day. Even the American Petroleum Institute disagreed at the time, saying "the only safe concentration for benzene is zero."[58] Devra Davis goes on to explain how manufacturers have continued to fight regulations on benzene up to the present time, even weakening enforcement efforts by the Occupational Safety and Health Administration (OSHA). OSHA oversight was only carried out in a serious way during a brief period in the 1970s when Dr. Eula Bingham was assistant secretary of labor and the head of OSHA under President Jimmy Carter.[59]

Alice Hamilton is also the foremother of an important contemporary environmental health campaign to end mercury pollution, especially from coal-fired utilities. Thousands of other medical and health professionals, citizens, and hundreds of American and international nongovernmental organizations (NGOs), many led by women, have been involved in these campaigns.[60] Hamilton had always been fascinated by mercury, or quicksilver, as it was often called in her day. She had seen cases of mercury poisoning in factories during the war, and mercury, like lead, had been a known health hazard since ancient times. She got her chance to study it and do something about its dangers when Dr. Henry S. Forbes, also concerned about mercury, offered to fund a Harvard study in 1923. Hamilton did the field investigations, going out to quicksilver mines and recovery plants in California at Santa Clara and New Idria. Feeling as if she had entered another country, she was surprised to find entirely Mexican or California Hispanic workers. They were exposed, especially in the recovery process, in a number of ways: through fumes, poor ventilation, and exposure to heated, liquid mercury, and more. Because such work could be rendered totally safe only if carried out in airtight chambers, an impossibility for such operations, she recommended that companies institute education about the dangers, frequent medical examinations, and meticulous maintenance and cleaning.

Her study then turned to the felt hat industry, a major one at the time. It used mercuric nitrate to hasten and improve the felting process. This involved workers pounding and steaming shaved rabbit fur, wetted by a solution of mercuric nitrate, which steadily released mercury into the air. The result, which she observed in felt factories in Philadelphia, New York, and Danbury, Connecticut, was "mad hatter's" disease, of *Alice in Wonderland* fame. Workers exhibited dry mouths, extreme irritability, and even dementia. As a result of her work and that of her Harvard colleagues, safer, nonpoisonous substitutes for mercury were found and introduced. The Mad Hatter became a thing of the past.[61] As with benzene, mercury's dangers became more and more widely known, thanks to Alice Hamilton.

Alice Hamilton's Protégée, Harriet Hardy

Alice Hamilton outlived Rachel Carson, but I have no direct evidence of their personal contact. However, in the small professional world of toxicology (the study of the effects of poison on humans), Alice Hamilton, in her retirement, and her colleague and friend, Dr. Harriet L. Hardy, who remained active until Carson's death, were likely known to her.

Harriet Hardy attended Wellesley College and the Cornell Medical School in New York City, and then became a resident at Philadelphia

General Hospital, one of the few hospitals in the early 1930s that would allow female physicians to serve. She went on to positions at the Northfield School for Girls, Radcliffe College, and then the Massachusetts Department of Health and the Massachusetts General Hospital in Boston. Hardy began work on the dangers of beryllium to human health, for which she later received recognition. After her first paper on this subject was published, she received a letter of congratulation from Alice Hamilton. They corresponded, and Dr. Alice, as Hardy called her, suggested they meet for lunch at a hotel in Boston. As Hardy tells it in her autobiography, "I had only seen her once before, when she had addressed a medical club to which I belonged at Wellesley College. Because I had a car at the time, I had been chosen to meet her and return her to the railway station. During these brief trips we managed to discuss a number of the world's problems. Some 20-odd years later we met in Boston, and with little introductory conversation, Dr. Alice invited me to be the junior author with her in updating her 1934 textbook, *Industrial Toxicology*."[62] Hardy protested that she lacked experience and was not competent for the job. Hamilton literally turned her one deaf ear toward Hardy and continued her recruitment pitch. When Hardy sought the advice of her superiors at the Department of Health and the Massachusetts General Hospital, they each told her that she must do what Dr. Hamilton asked. It would be good for her.

It was a wise choice. Hardy was eventually asked to join the faculty of the new Harvard School of Public Health, which had pioneered in industrial medicine and had hired Alice Hamilton. Over the years, the Hamilton and Hardy text was revised again and again, going to a fourth edition in 1982. Only then, because of age and a variety of ailments and tremors, did Harriet Hardy finally stop writing, though she remained an active speaker and mentor. She attributed the popularity of the text to a worldwide increase of interest in the environment and environmental disease. With a graciousness and recognition of mutual support from other women that characterized the network of female pioneers in nature study and conservation, public health and medicine, and social reform, Harriet Hardy also directly credited Alice Hamilton: "It is fair to say, however, that Dr. Hamilton set an early example of what a person with medical training, an observant eye, and social political sensitivity might accomplish."[63]

Harriet Hardy combined interests and helped to further the field of occupational and environmental medicine with innovative techniques, compassion for workers, and an ability to write and communicate to a variety of audiences. While at the Northfield School she dealt forthrightly with rape and venereal disease, instituted sex education, and taught preventive health education in anatomy and physiology—a sort of early precursor of the 1980s popular feminist book *Our Bodies, Ourselves*.[64] At Radcliffe and beyond, Hardy stressed broader learning and experience, encouraged the integration of disciplines, and generally sought to reform undergraduate and medical education. In a talk at the annual Massachusetts Medical Society meeting in 1975, for example, later published in a well-received article in the *New England Journal of Medicine*, she traced the history of industrial medicine "as reflected in the writings of poets, novelists, and men of public affairs, rather than in physicians' articles. In strong words (judged by letters and comments to me) I argued that most doctors as a result of narrow education in premedical and medical college years lacked knowledge of history, great writing, and philosophy. And as the number of medical specialty boards increases certain areas of medical knowledge of the physician, perhaps his interest in wider areas of his life becomes narrow."[65]

Harriet Hardy went on to study and publish on asbestosis, mesothelioma, black lung disease, radiation injury, and more. She was proud to work with labor unions, though critical of some of them, and disdained the "company physicians" she encountered who were willing to shave the truth for their employers. And like her mentor, Alice Hamilton, she was unconcerned with charges of radicalism. Her first studies of beryllium—which the Public Health Service previously had designated as harmless—discussed illnesses in the manufacturing of fluorescent lights. When she spoke and had her picture taken at the annual meeting of the various electrical workers' unions in 1947, Hardy explains that she was "considered very brave because Communists were present and the McCarthy Committee witch-hunts had begun. Years later, in 1970 . . . I was awarded a first Dr. Alice Hamilton Plaque by the unions. The presentation was made by an older, short man with tears in his eyes while he told of my courage by speaking in Boston."[66]

Like Alice Hamilton, Harriet Hardy continued to teach, talk, and later, correspond and mentor well into old age. She touched many, many leaders of the late-twentieth- and early-twenty-first-century battles over environmental exposures and human health. We can only assume that a thorough researcher and reader like Rachel Carson—with her interest in the chemical causes of cancer and having a wide network of friends, colleagues, and correspondents—was probably aware of both Hamilton and Hardy. A revealing clue is that Carson cites as one of her sources in *Silent Spring* a colleague, close friend, and mentor of Harriet Hardy, Dr. Oliver Cope of Harvard.[67]

Mary Amdur, Air Pollution, and Rachel Carson

As with many problems in protecting humans from environmental pollution, Alice Hamilton was limited, as was Ellen Swallow Richards before her, by the instruments at hand to detect, measure, and remove toxic exposures to pollutants. Industrial exposures were easier to observe, measure, and prevent because workers are often around large quantities or are directly exposed and show immediate, visible symptoms. Yet Alice Hamilton and others were well aware that even tiny amounts of lead or mercury or other chemicals could be dangerous. They were aware, too, that these contaminants could be and were spread through the air and water, often for long distances, until they reached humans either directly or through the food chain. But measuring tiny amounts, tracing their course through the winds, observing their accumulation in the body, and looking at their effect on human cells would have to wait for medical and technological advances. Rachel Carson, we know, needed to draw in the late 1950s on wide-ranging studies of the effects of DDT and other pesticides after they had been scattered by winds, found their way into food chains, and been absorbed into the systems of birds, animals, and humans. She was dealing with toxic air pollution.

If Alice Hamilton was the mother of industrial toxicology and the founder of environmental medicine, it is another of her protégées at the Harvard School of Public Health, Dr. Mary O. Amdur, who became the "mother of air pollution toxicology" at the time Rachel Carson was writing her early, best-selling works on the ocean. Amdur joined the Harvard faculty in 1949, when Harriet Hardy was there and Alice Hamilton, at eighty, was still mentoring female toxicologists like Amdur. Throughout her career, Amdur found both personal and professional support from, and carried out correspondence with, Alice Hamilton, Harriet Hardy, and Anna Baetjer, whom we will meet in chapter 3. Baetjer taught at Johns Hopkins while Rachel Carson was a graduate student and part-time instructor there.

Mary Amdur, Mother of Air Pollution Toxicology

Somewhat younger than Rachel Carson, Mary Amdur was also from the Pittsburgh area—born in 1922 in Donora, Pennsylvania, home to numerous mills and smelters and the site of the notorious killer smog of 1948. Like Devra Davis after her, also born in Donora, studying air pollution and disease became Mary Amdur's life's work. An exceptional student, she graduated from the University of Pittsburgh in 1943 with a bachelor's degree in chemistry and earned a PhD in biochemistry from Cornell University in just three years. Amdur started her career at the Massachusetts Eye and Ear Infirmary in Boston before joining the faculty at the Harvard School of Public Health in 1949 as an assistant professor to work with Philip Drinker, a leading pulmonologist and the inventor of the iron lung. Her task was to develop a test for lead in particulate matter, or soot, in the air. Without such measurements, manufacturers had been able to evade various efforts to control the use of and exposure to lead. Indeed, despite its well-known dangerous properties, lead was introduced into gasoline in the form of tetraethyl lead as an anti-engine knocking agent by the Ethyl Corporation. In 1925, a national commission looked at the effects of lead in from gasoline and declared it safe. The commission did so despite the impassioned pleas of Alice Hamilton, the leading authority on lead poisoning in the United States, who declared that even small quantities were unsafe and that alternatives must be found. Given massive industry lobbying and public relations campaigns over the years, leaded gasoline was not finally banned in the United States until 1978.

Modern work on toxic pollutants in soil, water, and air, along with Rachel Carson's ability to cite the long-range and persistent dangers of even small amounts of DDT, all grow out of the work of Ellen Richards, Alice Hamilton, and, then, Mary Amdur. It was Amdur who developed ways to measure lead in the air in small amounts; she also developed new animal tests to determine the adverse effects of very small exposures to toxins over time. First, she worked to measure the tiny amounts of pollutants whose undetectability had allowed the Ethyl Corporation to claim falsely that leaded gasoline was safe for gasoline workers and citizens. Amdur developed a novel hand pump to measure mere micrograms of lead in the air; it could be brought into the field whether in factories, garages, or repair shops. It was her ingenuity that confirmed the warnings of Alice Hamilton. Producing gasoline with lead in it, as Devra Davis puts it, was a horribly efficient way of dispersing this "insidious poison into the lungs, hearts and brains of millions."[68]

Next came the problem of the toxicological effects of small doses. The harm to humans from the Donora smog had indicated that low doses must have been involved. But it was only proven when Mary Amdur began an investigation of lung irritation from sulfuric acid in human lungs. The research was funded by the American Smelting and Refining Company (AS&R). The company had hoped to show that sulfuric acid was only a minor contributor to the Donora incident, and certainly *not* responsible for any deaths. Amdur worked under the supervision of Philip Drinker, whom the firm may have hoped would keep an eye on her research and results. They were sadly disappointed. Amdur studied both sulfur dioxide, the main AS&R emission, and the sulfuric acid that formed as a result; exactly the substances they did *not* want studied. At that time there were few studies on the effects on the cardiovascular system of inhaled pollutants. And they had only used lethality as the measure. If victims did not immediately die— no problem. Up until the 1980s, the main toxicological test, considered the gold standard, was called the LD50 test, short for a lethal dose where 50 percent of test animals die. Simply put, the dangerous levels of a chemical were determined after application to laboratory rats of enough of any toxic chemical so that half of them quickly keeled over.

It was Mary Amdur, working with her husband and their own funds, who invented a way to spray a fine sulfuric acid mist into a chamber of guinea pigs where humidity and air flow were tightly controlled. Guinea pigs, unlike rats, breathe through their mouths, not their noses, so that the mist enters deep into their lungs. This new research showed that damage to the lungs was caused by the irritants even when there were no lung spasms, then thought to be the real cause of damage. Mary Amdur went on to study the role of particulates, some not visible to the naked eye, in harming human health. She presented her findings at the AAAS meeting in 1953. Her smog research, carried out with tests on guinea pigs exposed to only small doses, showed there are dramatic effects on both guinea pigs and humans.[69]

Industry Retaliates against Mary Amdur

The story grows familiar, if more chilling, here. Amdur's research threatened entire industries. When she went to present her findings at the American Industrial Hygiene Association at its 1953 meeting in Chicago, Amdur was directly accosted. As she headed to give her paper on the ill effects of smog on the lungs of guinea pigs and humans, two big, tough guys in leather jackets got onto the elevator with her. They moved in close and looking directly at her said, "Hey, Mary, where you going? You are not going to present that paper, are you?" Amdur got off and delivered her paper regardless. But when she got back to Harvard, her superior, Philip Drinker, whose work was funded by industry, ordered her to take her name off the paper she had coauthored with him. He also told her to withdraw her paper from the prestigious British medical journal, the *Lancet*, where it had already been accepted. When she refused, Drinker fired her. The mother of smog research was unemployed. She received a note of condolence from Alice Hamilton, found years later carefully preserved in Amdur's papers. Eventually, Amdur was able to move to a different untenured position at the Harvard School of Public Health as a research associate under James Whittenberger. She raised all her own research funds and never rose above the rank of untenured associate professor at Harvard. In 1977, she moved again, still untenured, to MIT, which had, however tentatively, first admitted Ellen Swallow more than a century before. Despite her continued success at funding her own work and her growing fame, Amdur was stuck for twelve years in a nonfaculty position at MIT. At the age of sixty-seven, she again moved and funded herself at the Institute of Environmental Medicine at New York University in Tuxedo Park. Much later, John Spengler, who trained at Harvard, became dean of its School of Public Health. Responsible for many of the contemporary "green" initiatives there, Spengler writes of Mary Amdur: "I used to think that in science, it was sufficient to publish an important finding once. Amdur's work showed how foolish this notion was. At every step along the way, people tried to pull the rug out from under her. In fact, she got it right years before the rest of us. The world only caught up with her several decades later, by which time so many people had confirmed what she found that it could no longer be discounted."[70]

Spengler's words seem to describe Mary Amdur perfectly, but they also fit another intrepid woman scientist and writer who had been concerned with the harmful effects of pesticides as early as 1945 and of nuclear radiation during the atomic tests at Bikini in 1946. She, too, would need to write again and again, never gained real academic recognition, was viciously attacked by industry, and encountered overt and subtle gender bias throughout her life. Even after her greatest work, *Silent Spring*, reached millions, she would need to have her great work repeated, extended, and protected by those who came after her. John Spengler, who so admired Mary Amdur, could as well have been talking, of course, about Rachel Louise Carson.

3 ❧ CARSON AND HER SISTERS

Rachel Carson Did Not Act Alone

Within two weeks after it reached bookstores on September 27, 1962, Rachel Carson's *Silent Spring* reached the best-seller lists. It was the fourth blockbuster book in a row for the popular nature writer. But success was not guaranteed. Her colleague and friend Clarence Cottam, her Houghton Mifflin editor Paul Brooks, Carson herself, and others feared, with its grim message and serious science, that the "poison book" might not sell.[1] And all those involved with its publication knew it would be attacked harshly by the chemical industry and its paid scientific hacks and political allies. The assaults on *Silent Spring* and its author had begun as soon as a three-part abridged series in the *New Yorker* hit the newsstands on June 13, 1962. One letter writer called her a communist and a "peace-nut," while the Velsicol Corporation threatened legal action. In McCarthyite style, its general counsel claimed that there were "sinister influences" at work.[2]

Undoubtedly, controversy fueled the sales of *Silent Spring*. So did Carson's fine reputation as a writer. And *Silent Spring* had the huge benefit of selection by the Book-of-the-Month Club. But the outpouring of positive comments, the wide sales, and the impact of *Silent Spring* in scientific, conservationist, and Washington circles were not accidental. Neither were the defense of Rachel Carson and the warnings about the dangers of pesticides that were mounted in response to vicious attacks. Carson has been rightly credited with stimulating the modern environmental movement, bringing about long-neglected legislation, saving our most cherished and endangered birds, and with changing the way Americans view our relationship with nature. But, like the forgotten women naturalists and environmental health writers and leaders before her, she and her book did not suddenly start a movement *de novo*. Nor did she ignite a firestorm in a vacuum. Like Harriet Beecher Stowe, with whom she was compared, Rachel Carson and *Silent Spring* were the spark that lit long-smoldering tinder. Carson expanded the conservationist movement, shaped policy, alerted a wide public, added emotional and imaginative impact to accounts of the assault on the environment, and created a cogent, compelling scientific case documenting the havoc wrought by a rain of unregulated chemicals on America's backyards and birds. But the rousing response to a single book and its somewhat shy author had been building, largely offstage, away from the klieg lights, for a long time.

In 1958, as Rachel Carson slowly and slightly reluctantly decided to write *Silent Spring,* she drew on and increased her wide circle of friends, colleagues, conservation and civic organizations, and government officials who cared about the environment and public health. There was already a noticeable environmental movement, even though it had been reduced during World War II, become frustrated and somewhat futile during the early Cold War years of the Eisenhower administration, and was smaller and less coherent than today's. And much of this conservationist and environmental health movement had been founded, fueled, or fitted together by women. By 1962, as Carson continued to wrestle with difficult research material, family responsibilities, the death of her mother, a variety of ailments, and, finally, inoperable cancer, varied constituencies—from nature and scientific authors, to Audubon groups, the Sierra Club, Wilderness Society, National Women's and Garden Clubs, and researchers inside and out of government—became aware that an author with a gift for moving large numbers of Americans was about to present them with the opportunity of a lifetime. A perfect storm of events had also been gathering and was about to burst. Nuclear testing and fallout had already begun to upset and mobilize portions of the scientific and academic communities, and the concerned public. African Americans and their moderate and liberal white supporters were increasingly demanding protection and equal rights under the law; cancer-producing chemicals were discovered in the nation's Thanksgiving cranberries; deformed babies were born to mothers who had been prescribed the wonder morning-sickness drug thalidomide; and heavy-handed attempts to eradicate fire ants and gypsy moths with pesticides had gone awry. Domestic and social problems that had been shelved during the war could no longer be ignored. And then, in 1960, a charismatic and sympathetic young president was elected, along with a Democratic Congress. The torch, it seemed, had, indeed, been passed to a new generation. If, in politics and in life, timing is everything, Rachel Carson and *Silent Spring* arrived at just the right moment.

But such moments must be prepared for; Rachel Carson had been in preparation since she was a child in the seemingly distant Progressive Era. Carson worked easily and well with men; she had a number of male mentors, scientific and literary models, and good friends. But her deepest influences, relationships, networks and insights; her love of birds, of nature, and of science; her imagination and poetic gifts; her influential and political contacts; and her intimate personal support came from women.

Her closest relationship for much of her life was with her mother, Maria McLean Carson. Rachel Carson never married, and the two women lived together throughout the majority of her adult years. A formidable personality who was intellectually hungry after her own teaching career was cut short by rules against married women, Maria Carson encouraged her daughter's writing and career at every turn, just as she had tutored, trained, and groomed Rachel for a love of nature, science, and writing since childhood. With a distant and unproductive husband who had died young, it was with some sacrifice that Maria Carson sent her brilliant daughter to the Pennsylvania College for Women or PCW (now Chatham University) in 1925. An excellent small college, whose president and dean had attended Smith College, PCW had a strong liberal arts curriculum, including science, and an engaged faculty. But given the hobbled hopes for women at the time, few were expected to go on to graduate and professional study, or enter a career. But Rachel's upbringing, the desires and dreams of her mother, and her own observation of the disastrous domestic lives of her older siblings, led her to become part of a small, studious set of pre-professional undergraduate women. Everyone expected her to be an English major and writer. She had won silver and gold writing awards from *St. Nicholas* magazine at an early age, including her last youthful story, written when she was fourteen and published in the July 1922 issue of *St. Nicholas*. "My Favorite Recreation" shows how observant Rachel was of birds and nature and her romantic view of them. She was also an early, independent reader who deepened her interest in writing and her romantic view of nature by devouring the works, among others, of the wildly best-selling female novelist of her youth, Gene (Geneva) Stratton-Porter.[3]

Gene Stratton-Porter, the Best-Known Naturalist and Popular Novelist of Rachel's Youth

Gene Stratton-Porter was friends with Neltje Blanchan, whose husband, Frank N. Doubleday, started Doubleday and Page and published both women's immensely popular books. Stratton-Porter was a naturalist, wildlife photographer, and one of the first women to form a movie studio and production company. She was estimated to have some 50 million readers worldwide and wrote columns for popular magazines like *McCall's*. She wrote a number of nature books, but it seems that her novels, particularly *Freckles* (1904) and *A Girl of the Limberlost* (1909), originally published by Doubleday and Page, made a deep impression on the young Rachel.[4] They reflect the values of both the nature-study movement and of natural theology, a branch of American and English Protestant thought that emerged as scientific discovery continued to erode the inerrancy of the Bible; God can also be found through the appreciation of the intricacies and beauty of the created world.[5]

Both novels are set in the wooded wetlands and swamps of disappearing ecosystems in central Indiana. Freckles is a plucky, poor Hoosier boy with only one hand who badly needs money. He talks his way into a job patrolling and guarding the swamp from timber cutters and poachers of the animals there. Stratton-Porter's novels made such an impression on Rachel Carson that she alludes to them in *Silent Spring*, more than fifty years after their publication—and she assumes that her reference will be widely understood. She is regretting the intrusion of chemical sprayers into orchards and wishes someone would keep them out. "Does Indiana still raise any boys who roam through woods or fields and might even explore the margins of a river? If so, who guarded the poisoned area to keep out any who might wander in, in misguided search for unspoiled nature?"[6] Some of the parallels to Rachel's own life in *A Girl of the Limberlost* are uncanny and would have captivated her. The young heroine, Elnora Compton, like Rachel and Gene Stratton-Porter, is the youngest child by far. Both are doted upon, lack nearby friends, and are lonely. They grow up untrammeled, wandering happily in the woods that surround poor country farm homes that are slowly being encroached upon by towns and development. Nature is their friend and, in the case of Gene Stratton-Porter, it literally speaks to her through strange vibrations or emanations of some sort. Stratton-Porter even commented later in life that she understood the mysticism of her contemporary, the nature writer Mary Austin. Austin, who wrote about and fought for the desert wilderness of the West, had a vision of becoming a writer at five years old, while seated under a walnut tree.

In *A Girl of the Limberlost*, Elnora's widowed mother must contemplate lumbering or selling portions of their wooded property to make ends meet, as Maria Carson did. Elnora is extremely bright and studious, constantly reading and exploring nature, observing birds and bugs, and collecting rare butterflies. And like Rachel, she must make a long commute to high school where, because of her interests in nature and studying and her countrified ways and clothes, she does not at first fit in. Later, Elnora must work and teach biology in the city to earn money to go to college. Her mother shares her interests, studies her materials, rents a house near Elnora's school, and shares it with her. Throughout, it is clear that young Elnora is superior in learning and in values to her citified classmates because she has learned from the school of nature, not just books. She even has a female mentor, the Bird Woman, who knows all about natural history, shares her knowledge with Elnora, and collects specimens suitable for museums. And, as it turns out, life in the Limberlost swamp is better for Elnora and her wealthy, cultured fiancé, Philip Ammon, a lawyer from Chicago, who has broken off his engagement with the rich, beautiful, petulant Edith Carr. He rejects the predictable life of the members of his own class who come from mansions and sumptuous summer places on Mackinac Island. When he declares his admiration for Elnora, he reflects the values of the nature-study movement. "You are in the school of experience, and it has taught you to think, and given you a heart. God knows I envy the man who wins it. You have been in the college of the Limberlost all your life, and I never met a graduate from any other institution who could begin to compare with you in sanity, clarity and interesting knowledge."[7] Philip even declares that he wants a woman who can accept greater equality in marriage and a shared life of service to humanity with him. "I'm cured of wanting to swell in society. . . . I have no further use for lavishing myself on a beautiful, elegantly dressed creature who only thinks of self. . . . I want an understanding, deep as the lowest recess in my soul, with the woman I marry."[8]

Unlike her many natural history books, including several on birds, Gene Stratton-Porter's novels brought her fame and fortune. And, like Carson after her, Stratton-Porter did not shy away from environmentalism. She used the wealth from her works to support conservation of the Limberlost Swamp and other wetlands throughout Indiana and, at the end of her career, despite conservative leanings and conventional racially biased attitudes in her final novel, she wrote for *Outdoor America*, the new magazine of the conservationist Izaak Walton League. There she penned her final plea for the environment and building a movement. "If we do not want our land to dry up and blow away, we must replace at least part of our lost trees. We must save every brook and stream and lake . . . and those of us who see a vision and most keenly feel the need must furnish the motor power for those less responsive. Work must be done. It is the time for all of us to get together and in unison make a test of our strength."[9]

Rachel Carson's Female Mentors in College Shift Her Goals

We cannot know all that the bookish Rachel Carson read in her youth or later, but she was clearly familiar early on with the classics of British and American literature and later in life referred to a line from Tennyson's "Locksley Hall" as part of her earliest inclination toward the sea from landlocked Pennsylvania. As the literary critic John Elder has pointed out, John Keats's "La Belle Dame sans Merci" provides not only an epigraph for *Silent Spring*, but a framework for the book's critique of the whole enterprise of the human destruction of nature. Its conclusion, built around Robert Frost's "The Road Not Taken," avoids didacticism and leaves the sensitive reader to choose among the difficult, even enigmatic roads forward as humanity enters a new ecological future.[10] Rachel Carson headed off to college determined to be a writer, presumably one in the tradition of naturalist women like Cooper, Bailey, Wright, or Stratton-Porter. In her application to PCW, she spoke of her love of nature, said that "wild creatures are my friends," that she hoped to remain an "idealist," and, showing her ease with literature and her ambition, quoted Robert Browning's "a man's reach must exceed his grasp or what's a heaven for?" Once on campus, Carson wrote essays and poetry; joined the PCW newspaper, *The Arrow*; and contributed to the college literary magazine, *The Englicode*. By the spring semester of freshman year, she had become not only a star student, but also a friend of Grace Croff—a new assistant professor of English from Radcliffe with demanding standards, vitality, and imagination—sharing with her tea and long conversations on campus benches. Carson valued her teacher's opinion highly and was encouraged by the good grades that she received and Croff's comments on her ability to make "what might be a relatively technical subject very intelligible to the reader. Your use of incident and narrative is particularly good." But perhaps the most important comment from Croff to her prized pupil—who was already skipping social events to look at the bird collections in the Carnegie Museum of Natural History—was that "I have always felt that by using your imagination a little more you would limber your style."[11]

It was, however, a different circle of women who eventually moved the young Rachel Carson from English major and nature writer to biology major and scientist. Science jobs, whether in research or teaching, were scarce and difficult for women to acquire, so Carson's move to switch her major at PCW and her primary interest was a slow and somewhat difficult one. Among her undergraduate friends were Mary Frye, a biology major who planned to attend medical school, and Dorothy Thompson, a history major. Close friends by junior year in 1927, Carson wrote letters to both and shared the same biology class. It was here that she found her most important female mentor, one who would change her life and career, Mary Scott Skinker.

Mary Scott Skinker—Friend, Mentor, and Government Scientist

A stylish, attractive, and sophisticated professor of biology, Mary Scott Skinker, like Grace Croff, was popular with PCW undergraduates. Skinker had earned a master's degree at Columbia Teacher's College in New York and had been an outstanding high school teacher for some years. She also believed that young women could and should consider careers in science, an idea that was still quite advanced in the 1920s. Skinker planned to attend either Cornell or Johns Hopkins graduate schools for a doctorate and applied for a leave to carry out her plan after a summer at the Marine Biological Laboratory (MBL) in Woods Hole on Cape Cod. By her senior year, Rachel Carson and Mary Skinker had also become friends, as well as mentor and protégée.

In their frequent correspondence, Skinker raved about the opportunities for learning at the MBL and the welcoming atmosphere for women who wanted scientific careers. "To be engrossed in one's work is good form here and that's what I enjoy most." Carson clearly was fascinated by Skinker's descriptions. "It must be a biologist's paradise,"[12] she wrote to her premed friend, Mary Frye. Woods Hole had first been an experimental station at the seaside set up in 1881 by Spencer Fullerton Baird, the assistant secretary of the Smithsonian, who had helped and mentored Graceanna Lewis, Florence Merriam Bailey, and Martha Maxwell. When it opened in 1888, Ellen Sparrow Richards had been influential in raising funds, had trained its early leaders, and served on its board.[13] Thanks to Mary Skinker, Rachel Carson was about to be linked to a group of world-class scientists and important contacts, but also to enter a supportive atmosphere that still reflected the long legacy of its female founders. Carson was also so influenced by Skinker that she applied for early admission to Hopkins herself once it was clear that Skinker would be unable to return to teach at PCW. Carson was accepted but had to postpone going for financial reasons. Upon graduating summa cum laude from PCW in 1929, she then set off for Woods Hole with her friend Mary Frye.[14]

Carson was delighted with Woods Hole and her first real encounters with the ocean. She and Frye roomed together, studied, and prowled the beaches. But, most important for her must have been the introduction to a sophisticated set of nationally recognized biologists to whom she gained entrée. Carson's success and acceptance at Woods Hole by the leaders in her field—all in a relaxed social setting where women were treated fairly equally and expected to excel—must have quickly expanded her confidence and her horizons. She reveled in the first true research library she had ever used; was carefully guided in her research on the cranial nerve of turtles and other reptiles by R. P. Cowles, the biologist with whom she would study when she got to Hopkins; and, among her new contacts, was impressed by scientists from the U.S. Bureau of Fisheries.[15]

Rachel Carson and Women at Johns Hopkins

In the fall of 1929, when Carson finally got to Hopkins graduate school at the leafy Homewood campus in North Baltimore, she did well, but had to work extremely hard, given some poor preparation with the scientist who had replaced Mary Scott Skinker at PCW. Carson's research in the zoology department began on reptiles, but after difficulties in finding adequate specimens and innovative topics, she finally settled on a challenging thesis on the embryonic development of kidney function in catfish. But two important female support networks emerged while Carson carried out

FIGURE 3.1   Rachel Carson in the Pennsylvania
College for Women yearbook in 1928. Rachel
Carson Collection, Chatham University Archives,
Pittsburgh, PA.

her studies. Given her financial situation, she needed part-time work as
she pursued her two-year master's program. Carson began to teach undergraduate biology at Johns Hopkins, where she became fast friends with
Grace Lippy, who was already teaching while in grad school. Together
the two made a close and complementary pair, with Grace the outgoing,
enthusiastic teacher and Rachel the attentive and caring lab and research
leader. The pairing is reminiscent of other women who developed friendships and mutually supportive relationships: Florence Merriam and Fannie Hardy, who organized the Smith College Audubon club together;
Merriam Bailey and Olive Thorne Miller, who worked together and traveled to Utah and throughout the rugged West; Ellen Swallow Richards
and Bea Capen, who taught together at Girls High School in Boston; and
Alice Hamilton and Harriet Hardy, who collaborated at the Harvard School
of Public Health.

Grace Lippy's background was similar to Carson's. She grew up in rural Maryland, not too far from Pennsylvania, and, for financial reasons, lived at home for her first two years of college at Western Maryland College before she was able to transfer to Wilson College, a more demanding small women's college in Chambersburg, Pennsylvania. Lippy had completed her master's at Hopkins and was pursuing her doctorate in zoology. But she, too, needed to work to continue. Given an outstanding record, she was hired to teach zoology in the Hopkins summer school, the only woman to be appointed an instructor in zoology during the Depression years. When an opening came up, she gave up doctoral work and became assistant professor of zoology at Hood College in Frederick, Maryland, while continuing to teach summer school at Hopkins for the next four years alongside Carson.[16] Lippy became Carson's only close friend in graduate school and helped her get another part-time job, as Carson needed to pay off undergraduate loans and help her family. It was Grace Lippy who enthusiastically recommended Carson for an assistantship at the University of Maryland Dental and Pharmacy School in College Park, about thirty-five miles from Baltimore.[17] When Carson was about to begin her second year of study in the fall of 1930, it became clear that her original scholarship and funds did not sufficiently cover tuition. Because she needed to make more money, she desperately sought half-time work, even though by doing so she would be forced to reduce her commitment to grad school to half-time.

Rachel Carson managed by chance to find a lab assistant job that fall working on genetic experiments with fruit flies under Dr. Raymond Pearl at the Hopkins School of Hygiene and Public Health, where she found a wider women's network and new perspectives on the environment and health. There she would have been familiar with her boss's wife, Maud DeWitt Pearl, a biologist and managing editor of the journal *Human Biology*, as well as other female scientists at the Hopkins medical complex in East Baltimore. The Pearls were widely connected not only in the world of biology, but in politics and many other fields. In collaboration with, and assistance from, his wife, Raymond Pearl predicted population growth, corresponded with Margaret Sanger and advocated for birth control, challenged prevailing racist notions of eugenics and race, worked with the NAACP, wrote about food and nutrition, human genetics, smoking and longevity, the biology of the Great Lakes, and much, much more. It is impossible to know exactly how much Raymond and Maud Pearl and their circle affected the environmental health, intellectual, and political thinking of the young assistant, Rachel Carson. But it must have been substantially.[18]

Anna Baetjer, Chromium, and Cancer

Among those whom Rachel Carson also must have known at the Hopkins School of Hygiene and Public Health was Anna Baetjer, one of the foremost women pioneers in environmental and industrial health. A petite, wiry woman with "the energy of a dynamo," Baetjer was born in Baltimore in 1899 and graduated from Wellesley in 1920, echoing Carson's interests, with a dual major in English and zoology. Lively and witty, she was a popular teacher and mentor who helped found the field of environmental toxicology. The story is told that when she was at an advanced age, a male graduate student gasped to keep up with her always torrid pace as they toured an industrial plant. "I sure hope that when I'm your age, I'm in your shape," he panted. Not at all out of breath, Baetjer glanced at him and riposted wryly, "Why, how could you be? You're not in my shape now." She was also fearless, often entering prisoner lock-ups to measure ventilation shafts and air circulation and quality.[19]

But in the 1920s, given the overall attitude toward women and science, friends had discouraged Baetjer from even applying to graduate school. Nevertheless, like Ellen Swallow Richards, she persisted and talked to William Howell, professor of physiological hygiene, and later dean of the School of Public Health at Johns Hopkins in Baltimore. She must have impressed him, because she was accepted into the doctoral program, earning her Doctor of Science (ScD) four years later. Women researchers and faculty at the time were generally stuck in low-paying junior positions, whatever their merits. When Baetjer graduated with an exceptional record, Dean Howell offered her a job on staff, but only if she promised not to marry. Baetjer joined the faculty as an assistant professor in 1924, five years before Rachel Carson arrived. It was at a time, thanks to Ellen Richards, Alice Hamilton, and others, when interest was growing in state health departments and, at Hopkins, in how the home, the streets, and the factory were making people ill. As Baetjer recalled in an article in 1979 for the *Baltimore Sun*, "A lot of people talked about it, and there was a lot of speculation and guesses, but nobody really knew . . . what happened inside the body. It became clear we needed a lot of basic research."[20]

Ultimately, however, Baetjer's department was downgraded and folded into E. V. McCollum's Department of Chemical Hygiene. But with her energy, determination, and positive attitude, Anna Baetjer took advantage of her further reduced status to begin innovative and creative work. "I was left entirely alone," she said later. "Dr. McCollum said to me, 'I don't have any time, and as long as you stay in the budget, you can do anything you want.' "[21] With only limited resources, Baetjer turned out a steady stream of important and original research, first winning national recognition for her book-length study for the U.S. Army, which began to have greater numbers of women during World War II. In 1946, the National Research Council published her *Women in Industry: Their Health and Efficiency*.[22] Based on the observation and many studies of women at work, Baetjer's recommendations included adjusting machinery, equipment, environmental exposures, pregnancy and maternity leave, and working hours (what we might now call flex time) to allow for home responsibilities such as child and elder care. An impressive piece of research, synthesis, and clear presentation, Baetjer's book draws on and documents the work of a number of women scientists, most noticeably, Alice Hamilton.[23]

But it was Anna Baetjer's work in the 1940s on the connections between chromium and cancer that links her most closely to Rachel Carson and the impact of the environment on human health. Chromium was widely used by Baltimore's manufacturing firms, especially in chrome-plating car parts. But it exists in several different forms, is also a human micronutrient, and is very hard to study. Through painstaking research, however, Baetjer demonstrated that chromium (VI) is indeed a human carcinogen—causing lung and other cancers. By the 1950s, she was involved in helping to redesign the Fells Point Mutual Chemical Company in Baltimore to minimize workers' exposure to chrome dust through new technologies. Baetjer became a professor emerita in 1970 but remained active in research and consulting until shortly before her death in 1984. Outliving her contemporary and colleague Rachel Carson, she built on the attention that Carson brought to pesticides and human health, making national news in 1974 with a study that showed that 1.5 million workers in pesticide plants were at risk of cancer through exposure to arsenic. But in 1984, Baetjer, too, finally succumbed to breast cancer.[24]

Given her need to teach (and commute), her assistantship at the School of Public Health, and difficulties with her research for her master's thesis, Rachel Carson applied for and received permission to delay her degree completion from Hopkins by a year. As the 1932 academic year neared its end, she was named, with Grace Lippy, as one of several "investigators" admitted to work at the Woods Hole Laboratory, where they roomed together that summer. But there was little prospect for employment even after Carson had completed her master's in June and done further study at the MBL. Although she intended to pursue a PhD, the press of family, finances, years as a part-time instructor, and the difficulty in finding academic work finally forced her to give up her goal of earning a doctorate.

Yet, once again, a female mentor was instrumental in setting Rachel Carson on a course that would lead to her path-finding and popular work. Mary Scott Skinker had developed health problems while at Woods Hole, lacked funds, and, like Carson, had to delay doctoral work. She worked as a researcher for the government at the Zoological Division of the Department of Agriculture's Bureau of Animal Industry. It was Skinker, Linda Lear concludes, who had probably introduced Carson, in 1929, before she arrived at Johns Hopkins, to Elmer Higgins, head of the Bureau of Fisheries. Years later, in 1935, after she had prepared with Skinker to pass government zoology exams, Carson, at Skinker's suggestion, approached Higgins again. He happily offered her a part-time job in government as a researcher and writer. Thus, Carson began writing scripts for a 52-part series of public radio broadcasts to be called "Romance under the Waters," which none of the scientists at the bureau seemed willing or capable to do. The scripts were a hit, and Higgins's superiors were delighted. Carson was soon promoted to a full-time job in which her scientific training at top research institutions, her writing ability, strong work ethic, and imaginative and creative approach even to government brochures won her praise and promotions.[25]

Shirley Briggs and Carson's Legacy

After working successfully as a government bureaucrat for some years, Rachel Carson developed new female friends in government who became important sources of support, amusement, and networking while she worked in Secretary of the Interior Harold Ickes's Fish and Wildlife Service

(FWS), which had formally begun in 1940, combining the old Bureau of Fisheries and the Bureau of Biological Survey. Kay Howe was hired as an illustrator in 1944 and convinced her classmate and friend from the University of Iowa graduate school, Shirley Anne Briggs, to also seek employment as a scientist at the bureau. After Briggs was hired as an information specialist and illustrator in 1945, the three women soon were buddies at work, traveling companions, and part of a typical office clique that met in clandestine coffee klatches, as using hot plates or other office heating items was forbidden. With these pals as a personal and professional support network, Carson first showed her talent for the popular presentation of complex environmental subjects that might otherwise be simply technical or dull. She was asked to write a series of government booklets, Conservation in Action, to inform the public about the nation's wildlife refuges. These intensely researched, beautifully written pamphlets led to her traveling across the country, including a trip to the Bear River National Wildlife Refuge at the Great Salt Lake in Mormon, Utah, where she followed in the footsteps of Florence Merriam Bailey, Olive Thorne Miller, and Mary Austin, while foreshadowing Terry Tempest Williams.[26]

Shirley Briggs remained close to Carson long beyond their government days and after Kay Howe had moved west after marrying. Briggs would ultimately carry on Carson's work until the end of the twentieth century by founding the Rachel Carson Council, writing her own book on pesticides, lecturing and speaking where possible, and giving interviews about Carson and the struggle against chemical companies.[27] She shared many of Carson's interests, including literature and art, botany and books, and serious amateur ornithology. Carson had been an active birder her entire life, beginning a life list of species seen at an early age. While living outside of Baltimore during graduate school and part-time teaching at Hopkins and Maryland, she had regularly gone out to observe birds when time permitted. Like her friend, Shirley Briggs was an avid and knowledgeable birder who understood the habitat needs, ecological niches, and threats to the birds she watched and loved. She first brought Carson to Hawk Mountain, the premier promontory for observing migrating hawks and eagles that had been fought for and preserved by another pioneering woman environmentalist, Rosalie Edge. Each autumn, Hawk Mountain attracted admirers and amateur ornithologists from all over the country, including Briggs, Carson, and their friend, the famous bird artist and author Roger Tory Peterson.

They gloried in the gliding hawks and fall foliage. The decline in eagle and other populations at Hawk Mountain became a key piece of evidence in *Silent Spring.*[28]

Briggs also joined with Carson on work-related trips to refuges in North Carolina and elsewhere, and brought her along to meetings and outings of the Audubon Naturalist Society (ANS). It had been founded by Florence Merriam Bailey and others earlier in the century as the DC Audubon Club. Teamed with the slim, witty, and outgoing Briggs, Rachel Carson soon was a fixture with the inner circle of the ANS.[29] As with Woods Hole, which was both a scientific center and a relaxed retreat for the nation's top biologists, it is hard to overstate the influence and reach of the ANS. When it was founded, members like Florence Merriam Bailey were invited to Theodore Roosevelt's White House. By the 1940s, and especially when she was elected to its board of directors in 1948, the ANS linked Carson to influential conservation, government, and philanthropic circles that included the secretary of the Smithsonian Institution and the founder and president of the National Geographic Society.[30] Members like Roger Tory Peterson were revolutionizing birding and involving millions of Americans in the widespread recreation of bird watching, which they insisted was really a sport called birding. Peterson produced the first truly portable and complete field guide that used illustrations, simple distinguishing field marks and range maps all in one place. Published as the third edition of *A Field Guide to the Birds* in 1947, just in time for the suburban boom in peacetime America, it offered an expanded, thoroughly revised survey, with more illustrations than the first two editions that Peterson had put out in 1934 and 1939. Over the years, *Peterson's*, as it has become known, has sold millions of copies and stimulated numerous competitors and improved guides.[31] But, as we have seen, the modern obsession with, and love of, birds began nearly a century before Peterson with popular accounts, handbooks, and, finally, guides by women like Florence Merriam Bailey that appeared all the way into Peterson's day.

Early Warnings about Pesticide Dangers

Peterson and others in the ANS, like most Washingtonians, talked politics as well as birds and worried about how to preserve breeding, feeding, and migration grounds. And they also became aware of new threats to birds from the rapidly growing use of pesticides in American life. Even before World War II had ended, Shirley Briggs, Peterson, and a number of American scientists and journalists were already warning about the dangers of DDT. Because DDT undoubtedly saved many lives during the war as an easy and effective pesticide used in the control of typhus and malaria, its subtler and long-term dangers were ignored. After the war, with huge stocks and production capacity on its hands, the DuPont Corporation, the main producer of DDT, pushed its miracle chemical into the domestic market with a skillful and powerful advertising campaign and low prices that soon had DDT used as a panacea for any sort of pest. Other chemical companies that produced similar products, like dieldrin and chlordane, followed suit.[32] The 1960s stage was set for both *Silent Spring* and for congressional appearances by Rachel Carson and Roger Tory Peterson on the dangers of DDT and other pesticides.

But it was in the world of Washington insiders and birders that the warnings were first discussed and circulated. Carson was reading scientific papers on DDT in the 1940s as part of a circle that included Audubon Society and National Wildlife Federation leaders who were already attempting to influence policy. Keenly aware of the concerns of these and other environmental groups like the Wilderness Society, she appreciated the attempts to warn the public, such as an article for *Nature* by her friend Edwin Way Teale, a noted nature writer. She also carefully reviewed and knew the scientific literature emerging from the government's Patuxent Research Refuge in Maryland. In fact, articles on DDT by Elmer Higgins and Clarence Cottam, who remained staunch supporters of Carson, came to her desk for editing at the FWS in 1945. Increasingly interested in writing again and always looking to supplement her income and reach wider audiences, Rachel Carson had begun freelance magazine writing while in government. In July 1945, she submitted a proposal to the *Reader's Digest* for an article on the dangers of DDT. She wrote that everyone knows about DDT's ability to wipe out insect pests. But, she explained to *Digest* editor Harold Lynch, "the experiments at Patuxent have been planned to show what other effects DDT may have when it is applied to wide areas: what it will do to insects that are beneficial or even essential; how it may affect waterfowl, or birds that depend on insect food; whether it may upset the whole delicate balance of nature if unwisely used." Unfortunately, Lynch and *Reader's Digest* were not interested.[33]

Carson Draws on Women's Ocean Research

Rejected on pesticides, Rachel turned to other government publications and research at hand for her freelance writing. A sunken treasure's worth of information had been and was still being pulled out of the world's oceans, thanks in large part to the U.S. Navy's wartime need to understand all aspects of the sea in order to combat German U-boats and do battle with the Japanese navy in the Pacific. Wartime technological developments like sonar, radar, bathyscopes, and submarines allowed scientists to observe and map all aspects of ocean topography, currents, temperatures, and the previously unknown and amazing creatures who had long resided there out of sight from humans. Carson's writing about the sea drew on these scientific developments to which she was privy; her own observations of the ocean, its shoreline, and shore birds on her various trips; and her experience with some deep sea dives she insisted on taking at Woods Hole and in Florida.[34]

Encouraged by her bosses, Carson wrote more and more as a freelancer while working at the FWS. She had first been encouraged to send off work based on her government knowledge and research by Elmer Higgins. He had found her essay "The World of Waters" a delight, but a little too literary for government brochures. He suggested it was so good on its own merits that she should submit it to the venerable *Atlantic Monthly*. Higgins's enthusiasm greatly encouraged her; she began to offer other short pieces of journalism and, after about a year, a revised and polished version of "World of Waters" to the *Atlantic*. Acting Editor Edward Weeks was enthusiastic, as was the rest of the editorial staff. The piece was published as "Undersea." It immediately drew praise, comparisons to Jules Verne, and the attention of book publishers. Before too long, Rachel Carson had her first book contract and in 1941 published *Under the Sea-Wind* to critical acclaim. But sales were disappointing (fewer than two thousand copies) when, shortly after publication, the outbreak of World War II turned the nation's attention elsewhere.[35]

During the war, she continued to write for magazines and kept up with the latest in science. It is worth noting that the new scientific knowledge that Rachel Carson drew on for the ocean trilogy and her first best-seller in 1951, *The Sea around Us*, was the result of the scientific work of an unrecognized, yet impressively talented woman scientist, Lieutenant Commander Mary Sears. Sears was in charge of all oceanographic research for the U.S.

Navy during the war and played a leading role at Woods Hole long afterward.[36] Only a writer of Carson's ability was able to turn such otherwise arcane, technical information into best-selling books. It was her popularity as an ocean writer that positioned *Silent Spring* to receive major attention. Her achievements are remarkable, but they also draw on the legacy of work by women like Mary Sears and those before her at Woods Hole like Ellen Swallow Richards.

The Influence of Marie Rodell

Linda Lear has brilliantly helped us understand the crucial role of Rachel Carson's most intimate friend, Dorothy Freeman, in providing emotional support. But, as Carson was trying to publish a second book, it was the savvy of her literary agent, Marie Rodell, that mattered most. Carson had grown disillusioned with her first publisher, Simon & Schuster, and wanted to find a new, better, and more compatible one. Initially skeptical of using a literary agent, she finally listened to the advice of her friend Charles Alldredge that she needed one. In characteristic fashion, she set out in April and May 1948 to systematically interview potential agents he had suggested.

Marie Rodell, the one she chose, fit the bill; for many years she was Carson's main personal and professional confidante. A brilliant graduate of Vassar College from a comfortable and cultured family of assimilated Russian Jews, Rodell had studied at the Sorbonne, knew four languages, and was steeped in art, literature, and culture. She was widely traveled and took trips to follow her own passions of archeology and anthropology. Outgoing, stylish, witty, and occasionally bawdy, Marie Rodell had been the author of three fictional mysteries and was secretary of the Mystery Writers of America. She had done all this while working as an editor at the William Morrow Company in New York where she was well connected and respected in publishing circles. She had also published a textbook on the theory and technique of mystery writing that had gone through three editions, and she wrote often for women's magazines. Postwar layoffs in publishing drove her to setting up her own literary agency. A political progressive, Marie Rodell became the agent for Congresswoman Helen Gahagan Douglas, the former movie star who ran in 1950 for the U.S. Senate seat against Richard Nixon, who red-baited her as the "pink lady." Rodell also was the agent, starting late in 1957, for the first book, *Stride toward Freedom*, of the rising young civil rights leader Martin Luther King Jr. She corresponded with King, who shared manuscript drafts and letters with his close aides, Bayard Rustin and Stanley Levison. All were under surveillance by FBI Director J. Edgar Hoover. It may have been her work and friendship with Marie Rodell, in fact, which caused Rachel Carson herself to come under FBI surveillance.

Rodell was a talented and tough editor of Carson's drafts, cutting and reorganizing, or asking for more warmth and human interest. But she also had an astute strategic feel for publishing, publicity, and politics.[37] She not only negotiated successfully with Oxford University Press through some complicated dealings, but also helped Carson get a prestigious Saxton Fellowship that would allow her to take leave and return to Woods Hole to prepare for publications on preserving the fishery resources of the New England banks. Rodell accompanied her to Woods Hole in July 1949 and shared the excitement and inconveniences of ten days aboard the tossing research vessel *Albatross III*. On the trip Carson observed the incredible variety of sea creatures brought up by net and the latest oceanographic equipment, including a bathythermograph that recorded ocean tempera-tures and an echo-sounder that gave a picture of the undersea canyons off Georges Bank.[38] Marie Rodell also kept careful diaries that she shared with Carson, who then drew on them for three articles that appeared in the bul-letin of the National Academy of Sciences.

And it was Rodell who knew and introduced Carson to Paul Brooks, the editor in chief of Houghton Mifflin's general book division. Brooks was a courtly and sensitive Bostonian who was also friends with and the edi-tor of Roger Tory Peterson's field guide. Although Carson served on the board of the DC Audubon Society with Peterson, she had not met Brooks. He immediately hit it off with her and suggested she write a guide to the seashore. But, more important, Paul Brooks joined the widening circle of those who promoted and continued her work before and after her death. His biography of Carson, *The House of Life*, selected and included her lesser-known writing along with excerpts from her major works, and it put her life in the context of great writers. And, not surprisingly, it was Marie Rodell who suggested and encouraged Brooks to write that book. His later 1980 book, *Speaking for Nature*, published by the Sierra Club, remains a wonder-ful resource that places Rachel Carson in the line of great nature writers. It also includes a chapter on women, which helped to introduce and begin to revive the reputations of Florence Merriam Bailey and Mary Austin. But in his biography, Brooks mainly stresses Carson's role as a writer and is not interested in her scientific or conservation movement antecedents, or her personal and environmental support networks. To some degree, it is Paul Brooks's admiration for her writing and her quiet courage that inadvertently promotes the early stereotype that Rachel Carson was primarily a writer about nature, that she worked alone, and that she shows "the enduring ability of one dedicated individual to make an impact on society."[39]

After the immense success of *The Sea around Us* in 1951 and 1952, Rachel Carson was finally financially secure and able to leave her government job to concentrate on writing. By 1957, she built a larger, airier brick rambler as her home in Silver Spring; found, purchased, and expanded a cottage in the woods near the shore in Southland, Maine; and considered purchasing a tract of land to set aside as a woodland preserve through the Nature Conservancy. In Maine with Dorothy Freeman, she went on moonlit walks, explored tidal pools, and poured out her most intimate feelings. These included her anxieties, fears, and frustrations over becoming a surrogate parent for her five-year-old grandnephew, Roger, whom she took on as guardian and later adopted after her niece, Marjorie, died at age thirty-one.[40]

Roger was a difficult child, but, eventually, Carson came to enjoy showing him the joys of nature and taking him on numerous rambles along the shore and in the woods. There are delightful descriptions of Roger in her letters to Dorothy Freeman and others, as well as photos of him learning to explore nature. It is unclear what and how much children's nature literature Carson read as an adult, but we do know that one of her most interesting and original projects—part of her enduring legacy—was stimulated in part by the example of, and her close friendship with, the children's nature author Ada Govan. As early as 1945, she had been encouraged by another female friend, Sunnie Bleeker, to add a human dimension to proposed articles on bird banding that had been rejected by the *Saturday Evening Post* and *National Geographic*. Carson then turned her material into a human-interest piece, called "Bird Banding Is Their Hobby," which sold for $500 to *Holiday* magazine. It featured Dr. Oliver Austin, whom she had visited and observed one summer at Woods Hole at his bird sanctuary, Tern Island, near Cape Cod. She also met several other noted bird banders, including Ada Govan of the Woodland Bird Sanctuary in Lexington, Massachusetts. Nearly sixty when Rachel Carson first wrote to her, Govan was not only a dedicated bird bander; she was also a serious amateur ornithologist and the author of the acclaimed 1940 book *Wings at My Window*.[41]

Ada Govan and Rachel Carson's "Wonder Book"

An autobiographical account, *Wings at My Window*, tells how Govan, as a young housewife, had lost three infant children in a row and nearly lost her ten-year-old son to prolonged illness. She and her husband built a house in the woods in order to regain their emotional and physical health, but Ada suffered a fall there not long afterward that left her crippled. In the throes of deep depression and serious pain, Ada Govan was sitting by her window during a blizzard when she saw a chickadee desperately cling-ing to the sill. She began to feed it and other birds that came, discovering a reason to live herself. She turned to writing articles for *Nature* magazine about bird feeding and bird protection for families, "shut-ins," and "young mothers who wanted to train their children to grow up loving birds." When developers threatened her property, funds from her writing and from her wide circle of readers allowed her to save it and create the Woodland Bird Sanctuary. It was here that her observations and bird banding created her reputation with ornithologists. Govan was also following directly in the tra-dition of Florence Merriam Bailey, Olive Thorne Miller, and Neltje Blan-chan, who wrote for families and children, and Mabel Osgood Wright, who first turned a wooded home into a functioning sanctuary and wrote widely about it.[42] Govan, according to Linda Lear, became Rachel Carson's most intense emotional connection since Mary Scott Skinker, even though they met only a few times and built their relationship through correspondence. It was, Lear says, Ada Govan's chapter 12 of *Wings at My Window*, "Children into Bird Lovers," that suggested ways to teach youngsters how to love and protect birds and other wild creatures that particularly impressed Carson.[43]

In it, Govan talks about how her own son, David, came to love birds, and more important, to develop a set of spiritual values. She writes that boys can be the enemies of birds and describes their chasing and shooting at blue-birds armed with air rifles and sling shots. But with the right introduction to nature study and birds, even such boys can be made into lovers, not ene-mies of birds. It creates openness and bonds between mother and child and creates sympathy for all living things. "A boy who has seen a bird fly through snow and sleet straight to his protecting hands can never become a willful destroyer." Like her own son, and later her grandson, "such a boy who has grown up with a love of birds . . . has something that will enrich and sweeten his life for as long as life may last. It is a treasure that is free to all. To deprive a child of it is to stunt his spiritual growth and to hinder the full flowering of those traits of character that will make him a better man, a better citizen, and a lifelong friend and protector of his kinsfolk in feathers."[44] It is as if the original nature-study movement that so enamored Maria Carson in the late nineteenth and early twentieth centuries had continued unabated.

Indeed, Ada Govan describes knowing absolutely nothing about birds or writing about them before she became crippled. Then, in 1931, about a year after she had her epiphany with the chickadee at her window, she decided to write an article for *Nature* magazine that would be clearly aimed at beginners. She sought help and advice from Richard Westwood, the editor, who recommended binoculars and sent her bird books. It seems clear that among them probably was Olive Thorne Miller's 1899 *First Book of Birds,* whose tone and influence is reflected even in some of Govan's language. Like Miller, she talks about how observing live birds is more interesting than looking at or studying dead ones. Miller and Govan each talk about the allure of learning about the mysteries of bird migration, and Govan closely mimics the closing of Miller's introductory chapter: "One who goes into the field to watch and study their ways will be surprised to find how much like people they are. And after studying living birds, he will never want to kill them. It will seem to him almost like murder."[45] Similar sentiments and language are found, too, in Neltje Blanchan's 1907 *Birds Every Child Should Know.* Blanchan writes: "What does the study of birds do for the imagination, that high power possessed by humans alone that lifts them upward step by step into new realms of discovery and joy? If the thought of a tiny humming bird, a mere atom in the universe, migrating from New England to Central America will not stimulate a child's imagination, then all the tales of fairies and giants and beautiful princesses and wicked witches will not cause his sluggish fancy to roam. . . . Interest in bird life exercises the sympathies. . . . It is nature sympathy, the growth of the heart, not nature study, the training of the brain that does the most for us."[46]

Drawing on her own youth, her experience with her adopted grand-nephew, Roger, and many of the ideas from Ada Govan's *Wings at My Window,* and its antecedents from Olive Thorne Miller, Gene Stratton-Porter,

Neltje Blanchan, and perhaps other woman writers, Rachel Carson determined to write a book that would inspire wonder and imagination in children through exploring nature. Though often set aside for more pressing matters, her "wonder book" appeared first in 1956 as an article with photos, including one of Roger, in *Woman's Home Companion*, and then, after her death, in a beautiful edition called *The Sense of Wonder* with photos by Nick Kelsh.[47]

Marie Rodell, Marjorie Spock, and Strategy for *Silent Spring*

The story of how Carson's most important book, *Silent Spring*, got its title has been often told. She and her friends and colleagues not only worried about how an exposé of the hazards of miracle chemicals would sell; they wrestled with how to frame the issue and make it attractive and palatable to readers. Carson puzzled over numerous titles with Marie Rodell, Houghton Mifflin editor Paul Brooks, and others, but it was Brooks who first thought of calling the chapter on bird loss "Silent Spring." Rodell then became convinced that the entire book deserved that title. Given her own literary background, she also pointed out the lines from Keats's "La Belle Dame sans Merci" that became the epigraph and the frame for human hubris that shapes the central theme of *Silent Spring*. It was no longer a "poison book," a lament for birds, a public health tract, or a mere exposé of the chemical industry. It was an ecological elegy.[48]

Had there been no anticipated backlash to *Silent Spring*, there would have been no need for the lovely lyricism and romanticism of John Keats. But those who initially challenged local governments, health departments, and chemical companies were also, by and large, shrewdly political and pragmatic. Chief among them was Marjorie Spock—one of the original plaintiffs in the 1958 suit against pesticide spraying—who became Rachel Carson's friend, confidante, and connection to the wider world of influential Americans attacking DDT, dieldrin, and aerial dusting.

Carson wrote that it was her friend and fellow bird lover Olga Huckins who urged her to write the book that became *Silent Spring*. However, in making this claim, she was consciously shaping public sentiment and being politically shrewd in choosing how and whom to highlight as responsible for the work's origins. Carson had been mulling over the idea and the evidence well before her friend sent her touching descriptions and public protestations over the agonizing death of the robins in her bird sanctuary home. Olga Huckins was, of course, an actual aggrieved bird lover in Massachusetts. But the larger political purpose shared by Carson and Rodell was better served by presenting Huckins as an upset, upstanding, previously unknown citizen, not as an active member of the organized environmental movement. Focusing on Huckins functioned somewhat in the manner of the early, simple civil rights story of Rosa Parks and the Montgomery bus boycott. A tired, hard-working black woman needed to sit down and was needlessly harassed and arrested. It is an image that still circulates even though we know that Parks was deeply connected with and active in the NAACP; had attended the Highlander Center, a decidedly left-wing training school; and that other arrests for failing to move to the back of the bus had already been carried out. In this case, Olga Huckins was certainly a concerned birder, but she was also a sophisticated literary editor for the *Boston Post* who had previously reviewed *The Sea around Us* and was in nationwide correspondence with members of the Committee against Mass Poisoning, the main activist group opposed to pesticide spraying.[49]

In fact, Marie Rodell herself, always the PR strategist, made efforts a number of times not to have *Silent Spring* publically linked to the "freeloaders and the lunatic fringe," including then controversial types like J. R. Rodale, the organic guru, and his Rodale Press. Yet, it was Rodell who made the approach to Marjorie Spock, in Carson's behalf, in order to get background material. What Rachel Carson got in response to this request was a treasure trove of research materials, a close companion, and incredible connections.[50]

Marjorie Spock and Rachel Carson Join Forces

Marjorie Spock was one of the most important members of the 1950s conservation or environmental movement. She became close to Carson, introduced her to elite circles, and supplied a steady stream of research materials and advice. Yet, quite understandably, given the concerns of both Carson and Marie Rodell about avoiding as much as possible being immediately discredited and needlessly provoking public attacks, Spock operated mainly behind the scenes of *Silent Spring*. To this day, Marjorie Spock is yet another woman environmental advocate who remains insufficiently known. Seen in a contemporary light, she was far ahead of her time in lifestyle, litigiousness, and leftward leanings. The daughter of a prominent and wealthy attorney for Averill Harriman Sr.'s railroad, Marjorie was also sister to Dr. Benjamin Spock, the most noted American pediatrician and protester of the early antinuclear era. Although Ben Spock and his sister were close, there is little mention of him in the voluminous Carson literature, even though Rachel Carson herself was decidedly and actively antinuclear. She wrote for and regularly read the *Saturday Review*, whose editor, Norman Cousins, beginning in 1957, was one of the founders and chairman of the board of SANE: the Committee for a SANE Nuclear Policy, the largest and most influential American antinuclear organization.

Ben Spock was also an early, prominent member. A congenial man of broad interests, Ben Spock had, in his youth, been a devotee of Ellen Sparrow Richards euthenics movement, even teaching the subject at summer institutes held at Vassar College where Richards served on the board of trustees. But Spock split publically with SANE in the 1960s over its more moderate opposition to the Vietnam War. Ben Spock, a Brahmin, elite liberal, who like other SANE members, had been surveilled by local police and "Red Squads," publically supported resistance to the draft and was indicted in 1968 by the government for conspiracy. Although not convicted, he returned to SANE only decades later. Despite his social status and his standing as a best-selling baby book author, Ben Spock had become controversial.[51]

Spock's fate was much like that of the 1952 Nobel Peace Prize winner and antinuclear physician, Dr. Albert Schweitzer. Schweitzer was revered around the world as a humanitarian, but when he used his Nobel Prize speech to criticize open-air nuclear testing, the U.S. government put him under CIA and FBI surveillance, opened and monitored his mail, and tried to smear him.[52] In this atmosphere, much to Rachel Carson's credit, she remained staunchly antinuclear and admired Schweitzer and his philosophy of "reverence for life" so much that he is quoted in an epigraph in *Silent Spring*. But the centrality and influence of many controversial activists, even revered ones, is subtly and skillfully kept off center stage. In such a Cold War climate, there was good reason for underplaying Marjorie Spock's role in the enterprise. She was the life partner of Mary (Polly) Richards, an even wealthier, progressive member of New York society. A free spirit, Marjorie

Spock had not followed the predictable path for a talented young woman of her class by going to Smith College, as her parents had hoped. Instead, she went off to Switzerland, where she studied eurythmy, a holistic form of movement, with the noted Swiss practitioner, Rudolph Steiner. She had intended to introduce and teach Steiner's techniques in the United States, but this plan was postponed while she cared for her widowed mother and raised her younger sister. In the early 1950s, she bought a large property and house in Brookville, Long Island, with her friend Polly Richards. Spock was a proponent of vegetarianism and organic foods and had studied organic farming in Switzerland, and Richards suffered from serious digestive problems that required a pure, simple diet. Together, they established a two-acre organic garden, in itself a rather advanced notion when mainstream America ridiculed or reviled food faddists and vegetarians in strings of invective that often also included aspersions on one's Americanism or worse.

The two women were, of course, outraged when their home and organic garden were repeatedly sprayed by both state and federal planes with a mix of DDT and fuel oil—up to fourteen times in a single day. Marjorie Spock asked one of her neighbors, the noted ornithologist Robert Cushman Murphy, to help find a lawyer to bring suit. Although Murphy did not find a lawyer, he persuaded his friends to join the legal proceedings. Among them were J. P. Morgan's daughter, Jane Nichols, and Theodore Roosevelt's son Archibald.[53] As the suit neared trial, it was Spock who informed and led a wide network of environmental activists. It was, of course, classic organizing; she even used an early form of what we would now call social networking. As she later told Linda Lear, she had one of the earliest models of a thermofax machine, a smoky, smelly device, in her basement. From it, she sent out a daily update called "Today in Court" to a large list of people, among them Olga Huckins and the activist group the Committee against Mass Poisoning. This was a distinguished group of concerned naturalists that included Rachel's author friend Edwin Way Teale, Robert Cushman Murphy, and others. But it was only part of a much wider network. The DC Audubon Club, led by its president, Irston Barnes, had, for example, been opposing the U.S. Department of Agriculture's fire ant extermination program for some time. Rachel's friend and former colleague at Fish and Wildlife, Durwood Allen, was writing a piece for *Fisherman Magazine*; the Conservation Foundation had just issued a lengthy report on pesticides; the National Audubon Society had put out a bulletin critical of the fire ant program; and there were already investigations of wildlife damage by House and Senate committees and an official at the Department of Health, Education, and Welfare.[54] The Long Island lawsuit and others like it only broadened the growing movement that included Mrs. Thomas Waller of the State Garden Clubs—a group that had been active in conservation since the days of Ellen Sparrow Richards and Alice Hamilton.

Carson drew on these and even wider circles and, as an indefatigable researcher, found more and more evidence and conservationist allies. These included the League of Women Voters and the National Council of Women.[55] She collected "mountains of material" but was especially excited, according to Linda Lear, about coming upon a copy of the hearings of the Delaney Committee (the House Select Committee to Investigate the Use of Chemicals in Food and Cosmetics). It offered her strong, expert testimony and had set a precedent for regulation of harmful chemicals that industry was still working to undo while I was at Physicians for Social Responsibility some forty years later.[56] As Carson delved into materials provided by Marjorie Spock and others, she followed leads relentlessly, even working internationally, corresponding with Dutch scientists and finding, in a 1956 bulletin of the International Union for the Conservation of Nature and Natural Resources, the Albert Schweitzer quotation she ultimately used in *Silent Spring*.[57]

As Rachel Carson continued to widen and deepen her contacts, Marjorie Spock offered political advice as well as a steady stream of research leads and news, which she sent several times a week; Spock served as what Carson called her "chief clipping service." The two were, in effect, plotting an environmental campaign. Because fishermen would be concerned about the bioaccumulation of pesticides in the fish that they as well as osprey and pelicans ate, Spock urged her to make contact with the Sport Fishing Institute, adding an important constituency to their circle of supporters. Carson agreed that "hunters and fishermen are good allies and where I still have an occasional contact with writers and organizations of this sort left over from my days in Fish and Wildlife, I am now trying to sow more seeds of discontent."[58]

Wider Networks

Marjorie Spock continued to be at the center of a group of scientists and researchers whose material fed Carson's research and expanded her contacts, including the eminent entomologist E. O. Wilson of Harvard, who connected Carson to the current research of John George, whom she had been aware of in earlier years at the FWS. Wilson sent her his work on fire ants and a mimeographed copy of a Conservation Foundation report by George that gave a full account of the fire ant case, but had only been seen by the environmental community. Environmentalists had been fighting widespread aerial spraying of dieldrin and heptachlor (sometimes from World War II bombers), used in a fruitless U.S. Department of Agriculture effort to eradicate the fire ant, which, they argued, could be controlled by less dangerous means. Numerous reports of the deaths of insects, birds, and animals were piling up as the spraying continued despite mounting opposition.[59] Wilson encouraged Rachel to write: "The subject is a vital one and needs to be aired by a writer of your gifts and prestige."[60] When John George then joined the Physiological, Chemical, and Pesticide Branch at the Patuxent Refuge in the fall of 1958, Carson had access to the center of key government researchers on pesticides and wildlife. This in turn led to important contacts like George J. Wallace, an ornithologist at Michigan State University who saw the collapse of robin populations on campus and turned to research on pesticides and bird reproduction. Her chain of contacts also led her to Joseph Hickey at the University of Wisconsin at Madison. Hickey held the chair once held by Aldo Leopold. Hickey's important scientific work studied and confirmed the effects of pesticides in thinning eggshells in bird populations such as eagles, peregrines, pelicans, and ospreys that had been noticed and were cause for alarm nationwide.[61] Hickey himself had also been a young admirer of and mentored by one of the more flamboyant and adventuresome woman ornithologists and nature writers of the modern period, Fran Hamerstrom. Together, they collaborated in researching and fighting pesticides long after Carson's untimely death. Through Hickey, and probably through her own reputation in birding and scientific circles, Rachel Carson would have been aware of the former glamorous Boston socialite who ended up living, working, and roughing it in the wilds of Wisconsin.

Fran Hamerstrom and Ruth Scott

Rachel Carson's colleague Joe Hickey wrote a fond and admiring foreword to his mentor's lively 1984 account of her conservation efforts and trapping and banding of raptors (hawks, owls, and other birds of prey), *Birding with a Purpose: Of Raptors, Gabboons, and Other Creatures.* He says, "Fran may be the last living relic of the era of ornithologists who started out with a shotgun instead of a pair of binoculars."[62] He recounts her achievements in popular and scholarly writing and, with her husband, successful efforts to study the behavior and ecological problems of the endangered Greater Prairie Chicken, and their battles to preserve its dwindling habitat, described amusingly in her 1980 book, *Strictly for the Chickens.*[63] All in all, Fran Hamerstrom wrote eleven books and more than 150 scientific articles, won the Wildlife Society Award twice, and, in 1971, received the National Wildlife Federation Award for Distinguished Conservation with her husband, Frederick. She was the only woman ever to receive a graduate degree under the fabled Aldo Leopold. Leopold was her mentor and started Hamerstrom on her research on the prairie chicken and the ecology of its habitat in the Buena Vista and Leola Marshes of Wisconsin. In 1961, while Carson was in the midst of her widespread research for *Silent Spring* and in touch with Joe Hickey, Hamerstrom helped alert the public to the need for habitat preservation and formed the Society of Tympanuchus Cupido Pinnatus (Latin for prairie chicken). Some seven thousand volunteer wildlife observers helped collect the necessary data for this conservation project in rather rugged circumstances. Every one of this small nationwide army was hosted by Hamerstrom in her rustic home, an 1850s farmhouse without plumbing. Hamerstrom was active both before and after *Silent Spring.* She was a contributor, for example, to Hickey's important 1969 book, *Peregrine Falcon Populations: Their Biology and Decline,* published only five years after Carson's death. Together, Rachel Carson, Joe Hickey, and Fran Hamerstrom, along with many others, helped save the peregrine falcon.[64]

Another key contact and friend of Carson's who gave her entrée into even wider circles of conservationists was Ruth Jury Scott, whom she met in the summer of 1961 while at her home in Maine. Scott was going to attend the Audubon Camp in Maine with her husband and wrote to Carson, who invited them to visit. Ruth Scott, like Carson, was from Pittsburgh and state bird chairman of the Garden Club Federation of Pennsylvania. She had attended the Carnegie Institute of Technology in the 1930s, but illness forced her to stop. Instead, she became a landscape designer and avid naturalist who designed and built a house overlooking Powers Run above the Allegheny River. It was only five miles away from Rachel's girlhood home in Springdale, an area that Scott was determined to save in its natural state. With an outgoing personality that complemented Carson's more reserved side, the two quickly hit it off. Ruth Scott was starting the Roadside Vegetation Management Project, designed to teach people about the dangers of pesticides and the positive uses of ecological vegetation management. By 1961, she had also become a central and effective leader in the Federation of State Garden Clubs and was on its National Bird Committee. She was also heavily involved in national and state Audubon Society efforts, as well as the Nature Conservancy. As their friendship developed, the two traded advice on how to advance things politically in the Garden Clubs, how to emphasize the dangers of pesticides more forcefully in Audubon reports, and more.[65]

Rachel Carson, Marie Rodell, and an Organized PR Campaign

As publication of *Silent Spring* drew closer, Carson cautiously weighed in with selected speeches and responses to chemical company denials of danger and their PR efforts. One of her letters to the editor in the *Washington Post* drew a private, favorable response from its publisher, Agnes Meyer, who wrote, "At this moment the gingko trees in front of my house all have a sign for motorists to beware because the trees are going to be sprayed in a few days. I wish the birds could read." Meyer continued to correspond with Carson and then to call by phone to offer encouragement and support. Thus was born the idea of an important luncheon launch of *Silent Spring* in the nation's capital once it was published.[66] Soon, that core idea blossomed into planning for a widespread publicity and organizing campaign for the book, instigated mainly by Marie Rodell. Working with Carson and her friend the publicist Charles Alldredge, Rodell plotted a nationwide grassroots and elite strategy to gain widespread support and stimulate sales for the book. They all knew that industry attacks and possible lawsuits would be forthcoming.[67] But Marjorie Spock more astutely reminded Carson of the serious damage and political difficulties such criticism could bring. Returning from a panel that Carson had decided not to address in order to avoid premature attention to her work, Spock reported that someone had announced that Rachel Carson was working on a book on pesticides, the opponents were organized, and "the government boys" who support the pesticide industry "were all still there" to hear the leak.[68] Thus, support and contacts with environmental groups became increasingly important. Carson relied heavily, for example, on the National Audubon Society's Harold Peters, a research biologist who had previously been at the USDA's Bureau of Entomology and a flyway biologist for the Fish and Wildlife Service. Peters was an expert on the southern states and the fire ant eradication program. He was also a passionate ornithologist who "knew everyone in wildlife biology, from Ivy League scholars to garden club leaders . . . and most of the field scientists for both USDA and FWS."[69]

Their campaign resembled one decades later that catapulted Al Gore's movie, *An Inconvenient Truth*, to record viewing levels for a documentary film, especially one on the equally dismal and dreary subject of climate change science. Building opening weekend attendance, giving away DVDs, arranging home and church viewing parties, gaining advance praise, and alerting key policy and political leaders for Gore's rollout were all part of a strategy to which national environmental organizations were central.[70] For *Silent Spring*, Carson, Rodell, and Alldredge discussed sending advance proofs of the *New Yorker* serialization that would precede the book to key players; they also brainstormed over whom to invite to a Washington launch. Both Rodell and Carson were keenly aware that an old-line Boston publishing firm like Houghton Mifflin would be wary of any promotion that looked like political lobbying and unlikely to distribute large numbers of advance copies of the book in Washington. As Lear reports, Marie Rodell and Alldredge had a much more aggressive campaign in mind. "They wanted to put a copy of *Silent Spring* in the hands of policy makers, cabinet secretaries, White House staff, congressional committee chairmen and their staffs, conservation organization directors, and women's organization leaders."[71] Paul Brooks and the Houghton Mifflin publicist, Anne Ford, were skeptical about a Washington launch and the huge number of advance copies. But Rodell was convinced that Carson would be attacked as a "crackpot and subversive"; she would need all the respectable and active support she could get. Ford feared that a Washington lunch would alert industry lobbyists, but Rodell brushed her objections aside, rightly pointing out that "we might as well get our ammunition ready ahead" as industry would be on full alert as soon as the *New Yorker* serialization appeared. Putting the book early into the hands of the leaders of groups like the Garden Club and the

League of Women Voters was also critical, Rodell argued, finally gaining the grudging agreement of Ford and Paul Brooks.[72]

The highly strategic and political plan began with a luncheon at Agnes Meyer's house that included on its invitation list former Secretary of Labor Frances Perkins, Senator Maurine Neuberger, Congresswoman Leonor Sullivan, Katherine Bain of the Children's Bureau, and the heads of the League of Women Voters, the National Federation of Women's Clubs, the National Council of Jewish Women, the Garden Clubs of America, the American Association of University Women, and the National Council of Women of the United States. This women's power lunch led to engagements to speak at the Waldorf Astoria in New York for the National Council of Women and before the Committee on Economic and Social Affairs of the AAUW. Rachel Carson also then received an invitation to, and attended, the White House Conference on Conservation as a distinguished guest. She was accompanied by Ruth Scott and her friend, Interior Department publicist Nicki Wilson. Always adept at social and political events, Ruth Scott made sure that Carson met Secretary of the Interior Stewart Udall, who had convened the conference at President Kennedy's request. She also met with Justice William O. Douglas, Sierra Club director David Brower, and Howard Zahniser of the Wilderness Society. Afterward, she began meeting regularly with Secretary Udall's senior staffer, Paul Knight, who had been assigned to follow reaction to *Silent Spring* in order to gain political information and develop policy ideas for Udall. Then Marie Rodell reported that *Silent Spring* had been selected for the Book-of the-Month Club and that the National Audubon Society, whose president, Carl Buchheister, had gotten one of the advance copies, was going to print a two-part series excerpted from *Silent Spring* in *Audubon* magazine. Consumer's Union soon followed with a request to create a special soft cover edition of forty thousand copies of *Silent Spring* for their members. And congressional support also began to appear from Representative John Lindsay of New York, who read portions into the *Congressional Record,* and from Senator William Proxmire of Wisconsin.[73]

Justice Douglas had written the stinging dissent from the Supreme Court decision that ruled against Marjorie Spock and other pesticide plaintiffs. He was ready to speak out further, and he penned a clearly memorable blurb for Houghton Mifflin, calling *Silent Spring* "the most revolutionary book since *Uncle Tom's Cabin.*" Then he wrote a favorable report on *Silent Spring* for the Book-of the Month Club *News,* saying "This book is the most important chronicle of this century for the human race." The initial run of this Book-of-the Month selection was 150,000 copies. And, in an unusual move, its board president, Harry Scherman, urged members to read the book, saying, "The portentous problem it presents—of worldwide poisoning of all forms of life, including human—is in the same dread category as worldwide nuclear warfare."[74]

Meanwhile, Paul Knight reported that Secretary Udall had become a strong advocate for pesticide reform in President Kennedy's cabinet, that a number of congressional committees were interested in hearings, and that Kennedy's science adviser, Dr. Jerome Wiesner, had called together the concerned agencies and appointed a task force to review all departmental policies on pesticides and report to the president.[75] Asked at a press conference whether his administration would take a closer look at pesticides, President Kennedy replied on national television, "Yes, and I know that they already are. I think particularly, of course, since Miss Carson's book, but they are examining the matter."[76]

Publicity and political support for *Silent Spring,* assembled shrewdly into a model environmental organizing and media campaign by Rachel Carson, Marie Rodell, Ruth Scott, and their allies, was reaching critical mass. Requests for television interviews and other appearances poured in to Marie Rodell. She weighed these tactically in the light of public and political reaction, as she personally supervised how everything was handled by Houghton Mifflin, including fan mail, speaking engagements, advance sales, and syndications. Just as Carson had been careful with her scientific references in order to be as unassailable as possible, so did Rodell make sure that support and publicity for *Silent Spring* not be associated with "free-loaders and the lunatic fringe—the ones with something to sell." She took pains to ensure that ads did not appear in then controversial outlets like Rodale's organic gardening or Health Guild publications.[77]

The campaign culminated with the release of the report on pesticides by President Kennedy's task force. It said the evidence showed that pesticides were, indeed, dangerous and that protective legislation should be enacted. When Eric Sevareid summed up the impact of Rachel Carson in a special *CBS Reports* called "The Verdict on the Silent Spring of Rachel Carson," he showed old clips of her reasoned, articulate, yet strong presentation of the facts in her earlier debate with a representative of industry. Sevareid referred to the report of the president's panel, released that evening, and said it showed that Carson had accomplished her goals of alerting the public and "of lighting a fire under the government." Before a huge national audience far beyond that of the *New Yorker*, Rachel Carson had been vindicated.[78] It was an important moment, because industry had threatened to sue Houghton Mifflin and the Audubon Society, had sent out paid scientific surrogates from its camp, and because Carson had been variously denounced by critics as a spinster, unscientific, a pro-communist, and more. She withstood it all calmly and bravely and spoke back directly and winningly in public. But planning and publicity, long before *Silent Spring* appeared, had been crucial. In private, Carson, Marie Rodell, and others kept weighing the impact of her words and appearances on the public and policymaker. When congressional hearings on pesticides began, Rachel Carson entered to television cameras, the assembled press, and sympathetic senators. Almost to a man they crooned to this new Harriet Beecher Stowe, "so you are the little lady who started it all." She had crossed a rare threshold into history. The earlier conservation movement, in which women had been so formative, had been reenergized, emerging now as the environment movement. Rachel Carson had triumphed. But she was not alone.

4 ❧ RACHEL CARSON, TERRY TEMPEST WILLIAMS, AND ECOLOGICAL EMPATHY

June 10, 1963, is rapidly growing hotter and more humid at American University's commencement in Washington, the sort of day that calls up the old jokes about swamps and hazardous duty pay for time served in some tropical capital. As the familiar, flat Boston tones begin, it is soon evident to faculty, students, and parents alike that they are witnessing history. Hand occasionally jabbing, as always, at the air, President John F. Kennedy calls for an end to the Cold War, an end to endless enmity with the Soviet Union. All of civilization, indeed the planet, is threatened by nuclear weapons and nuclear warfare. The endless rounds of open-air nuclear testing are harming humans right now. All of it must stop. Kennedy is finally and forcefully responding to a decade-long citizen campaign to end nuclear testing and to the most recent Soviet moratorium. And he is eloquent about it, rising to a carefully penned peroration: "We all love our children. We all breathe the same air. We are all mortal."

Kennedy's American University speech led within fifty-five days to the signing with the Soviet Union of the Limited Nuclear Test Ban Treaty. It put a halt for all time to open-air nuclear testing. But it reflects more than that. It is a monument to the world's first global environmental health campaign. President Kennedy properly gets the credit in the history books, with perhaps a note that his plea was written with his special counsel and confidante, the conscientious objector Ted Sorensen. And the giants of the antinuclear movement sometimes get a brief nod—the Nobel Peace Prize winners Albert Schweitzer and Linus Pauling, or leaders like Jawaharlal Nehru,

Lord Bertrand Russell, Norman Cousins, or Benjamin Spock. Rarer still is mention of the broader movement and its organizations like SANE; even less so the role of women like those in the Women's International League for Peace and Freedom (in which Dr. Alice Hamilton was active) or the newly formed Women Strike for Peace led by Dagmar Wilson.[1]

Yet, it is even more surprising to discover, and still little noticed, that Rachel Carson was a supporter of the campaign to end nuclear testing and rid the world of nuclear weapons. But it should not be if we see efforts to stop nuclear testing as a global environmental health campaign battling transboundary pollution—dust and debris that just happened to be dangerously radioactive. And so Kennedy's words that sweltering day at graduation must have had special meaning for Carson. But, in our time, Kennedy's great address is not seen as an environmental one. With a few exceptions, the modern environmental, health, and antinuclear movements remain separate, siloed, and seriously less effective because of it. Rachel Carson would have had none of these divisions. She portrayed the perils of pesticides and the problem of nuclear weapons as twin signals of human hubris. So, too, does writer and advocate Terry Tempest Williams, who sees our failure to empathize with others, and with other species, as at the heart of our current ecological and ethical crisis.

Rachel Carson's farsighted views also led her to see as interrelated a number of other causes and American reform campaigns. Despite Marie Rodell's concern that advance publicity not associate *Silent Spring* with the "lunatic fringe," this decision was more tactical than principled. Carson was an early and strong supporter of animal welfare and rights and spoke out for them.[2] And Carson's close association with Marjorie Spock and her partner, Polly Richards, with their organic, vegetarian, and alternative lifestyle, though kept from the spotlight, surely fits the caricatures drawn by her enemies. If Carson and her allies were constantly judging public and political response to their pronouncements, it is even more impressive to see her support for the antinuclear campaign in the midst of the Cold War, the Berlin crisis of 1961, the Cuban missile crisis of 1962, and the battle for an end to open-air nuclear testing that had been building in the United States since 1954. On March 1, 1954, Castle Bravo, the code name for an American test of a large thermonuclear device, had gone badly awry in the Pacific. It showered the Japanese fishing vessel, the *Lucky Dragon*, with radioactive fallout, killing the radio operator, giving the rest of the crew serious radiation poisoning, causing panic in the Japanese tuna industry, and provoking worldwide outrage; protest began in Japan, then spread to the rest of Asia, Africa, and, finally, Europe and on to the United States.[3]

Rachel Carson had opposed nuclear tests from their earliest 1946 origins in the atolls of the Pacific Ocean immediately after the war and the destruction of Hiroshima and Nagasaki. Interestingly, she learned about them through biologist Roger Revelle, later better known as the scientist who, in 1958, had begun a project atop the Mauna Loa volcano in Hawaii to measure the concentrations of carbon dioxide in the atmosphere; he is the professor who, in 1966, introduced and interested a young Harvard undergraduate and senator's son named Al Gore to the science of global warming. Carson had met Revelle during World War II when he was with the Bureau of Naval Research. In 1946, he became chairman of the committee studying the biological effects of radiation from Operation Crossroads whose two atomic weapons were detonated at Bikini Atoll (hence the origins of the hot new two-piece bathing suit, the bikini). The atomic bombs, one dropped from a B-29 bomber, the other exploded underwater, were deployed against surplus navy ships to show their power at the start of the Cold War and to quell growing fears and criticisms that their use might destroy the world. Useful information, if atomic bombs were ever to be used in the Cold War, was also sought on the effects of the bomb and radiation on life forms in and around the tiny islands and on test animals aboard the ships. Carson had edited the Fish and Wildlife Service (FWS) biological survey on the islands before the nuclear explosions and got to read Revelle's reports after the atomic devices were used. Like many other government botanists and biologists, she opposed the Bikini tests.[4]

By August 1952, Carson had gained a national reputation with the success of *The Sea around Us*. Roger Revelle invited her to join him as the historian for a four-month expedition called Operation Capricorn to the South Pacific.[5] It was designed to explore the western Marshall Islands. Revelle hoped that if she could not make the entire trip, she would, at a minimum, be able to join the segment of the expedition that included Kwajalein Island. It would start in November and return to San Diego in February 1953. Carson was eager to make the journey and learn about the South Seas firsthand. But to her disappointment, family responsibilities kept her away. She must, however, have paid particular attention and have been disturbed to discover

that the Marshall Islands and their residents were subjected to unexpectedly high levels of nuclear fallout during the 1954 test.[6]

When the Committee for a SANE Nuclear Policy was started in New
York in 1957 in order to attract a broader group of prominent supporters beyond the predictable group of Quakers and other pacifists, Rachel
Carson would have been immediately interested, even before meeting
Marjorie Spock. She regularly read the well-respected *Saturday Review of
Literature*, having been particularly pleased by a favorable review of her
first book, *Under the Sea-Wind*, by the renowned biologist William Beebe
in 1941. After excerpts from her soon to become best-selling book, *The Sea
around Us*, were released in a *New Yorker* serialization on June, 2, 1951, her
picture appeared on the July 7, 1951, cover of the *Saturday Review*.[7] Carson
also came to know its editor, Norman Cousins, who had been one of the
judges who awarded her the 1950 Westinghouse Science Writing Award and
had published William O. Douglas's dissent in the Supreme Court pesticide
spraying case.[8] Cousins was one of the key founders of SANE and its first
chairman of the board. He was also one of the first and very few prominent
Americans to write about and oppose nuclear weapons and the bombing of
Hiroshima. His immediate reaction was the August 6, 1945, editorial in the
Saturday Review of Literature, "Modern Man Is Obsolete." Cousins quickly
turned his antinuclear essay into the 1946 book, *Modern Man Is Obsolete*; it
went through fourteen editions and was read by millions.[9]

The early SANE also included the famous American pediatrician and
best-selling author, Dr. Benjamin Spock. His sister, Marjorie Spock, was,
as we have seen, one of Carson's closest friends and advisers. Membership
lists for SANE from the early 1960s are no longer available, but we do know
that in her final speech in 1963, "The Pollution of the Environment," Carson
cited the antinuclear Committee on Nuclear Information in St. Louis and
was friends with scientist Barry Commoner, who worked closely with the
Nobel Peace Prize winner Linus Pauling, also deemed subversive.[10]

By the time that open-air nuclear testing and the threat of nuclear war
were major issues in the late fifties and early sixties, Carson was concerned
not only about nuclear fallout, but radioactive wastes in general. She
remained aware of the latest developments in oceanography as an elected
member of the board of the Woods Hole Marine Biological Laboratory and
corresponded with the scientists there.[11] Not only would she have known
of the fate of the Marshall Islands and other test sites, as an expert on the

oceans, but she also became opposed to the widespread practice of dumping nuclear wastes into the supposed safety and security of the sea. In 1961, she raced to complete a new, revised edition of her decade-old best seller, *The Sea around Us*. Its preface, which was published in 1962, is an excellent example of what has come to be known among environmental health advocates as "the precautionary principle." Carson lays out a careful case of the potential dangers to both humans and other creatures through the dumping at sea of radioactive wastes encased in concrete casks. The nuclear caskets will have not reached eternal rest; they will be buffeted and ultimately broken by a dynamic ocean bottom. We know that radioactivity is dangerous, she argues; we know, too, that even concrete casks do not simply sink into some silent, immutable grave. Before the research of World War II and the Cold War, the ocean depths were seen as lifeless. Carson reminds readers that the oceans are in actuality full of life forms and are an active, evolving entity. "Other photographs," she writes, "give fresh evidence of life at great depths. Tracks and trails cross the sea floor and the bottom is studded with small cones built by unknown forms of life or with holes inhabited by small burrowers."[12]

The preface to the second edition of *The Sea around Us* concludes with a lucid synopsis of the interaction of life forms and their environment; the distribution, absorption, and bioaccumulation of toxic materials in organisms, the latest findings about radiation, and a clear warning of the dangers of dumping radioactive wastes into the ocean. Carson picks up her antinuclear theme even more strongly in her final speech before she died, "The Pollution of Our Environment." It was presented in October 1963 to 1,500 physicians and health-care providers at the annual symposium in San Francisco of the Kaiser Foundation Hospitals and the Permanente Medical Group; it contains a detailed scientific critique of nuclear testing, fallout, and radioactive wastes. One of the best scientific studies of the dangers of fallout, she says, is the recent discovery of highly elevated levels of strontium-90 and cesium-137 in Alaskan Eskimos and Scandinavian Lapps. Not because nuclear fallout fell especially heavily in the north. It did not. "The reason," she says, "is that these native peoples occupy a terminal position in a unique food chain. This begins with the lichens of the arctic tundras; it continues through the bones and the flesh of the caribou and the reindeer, and at last ends in the bodies of the natives, who depend heavily on these animals for meat." The radioisotopes in nuclear fallout are first absorbed and concentrated by the slow-growing lichen, or arctic moss. By the time they are found in Eskimos, as measured by scientists at Hanford in the United States, the concentration of radioisotopes is up to eighty times higher than that of Americans near the Hanford nuclear site in Washington State where the plutonium for the Nagasaki bomb was produced.[13]

Rachel Carson kept up with the latest developments in the scientific literature, including nuclear studies. She speaks knowledgeably in her last speech about why, in 1963, it has just been discovered that radioactive iodine, I-131, is a health hazard, especially for thyroid disease and thyroid cancer. The government had asserted all along that it was not. The problem has been, she says, that milk samples have been taken from national brands, mixed from sources all over the country so that any radioisotopes present are diluted. Nationwide local sampling reveals "hot spots" resulting from intense local fallout, which is then absorbed into the food chain and consumed. Measurements had also focused on external sources of radioactivity. Iodine-131, with a short half-life of eight days, was believed to be so quickly rendered harmless as to prevent hazards to humans. But the risk to citizens is from the I-131 absorbed internally and magnified by bioaccumulation in local milk. This is best understood, Carson explains, by the five recent atomic tests at the Nevada site in July 1962, whose fallout drifted over Utah. By examining individual samples of local milk, the Utah State Department of Health and the Atomic Energy Commission's Division of Biology and Medicine had decided that the fallout and consequent exposures created a hazardous situation.[14]

Rachel Carson's evocative and accurate writing about pesticides and human health in *Silent Spring* made an immediate difference. The same cannot be said for her sensible, sensitive, and scientific speeches and writing about the dangers of the nuclear age. Carson rightfully feared that the end of open-air nuclear testing might ease public concerns, seeming to solve the problem when it had not. Within less than a year, both John F. Kennedy and she were dead. Kennedy had wanted a total test ban, but the Joint Chiefs of Staff objected and stoked the fears of their congressional and conservative allies. Only a partial test ban was concluded. American nuclear testing proceeded apace underground, as did the production and deployment of nuclear weapons. The American environmental movement broadened and grew. Air pollution, burning rivers, and the wrecking and withering of wild places were visible; the threat of nuclear radiation or nuclear war was not.

The antinuclear movement continued, much reduced, a mere remnant of its glory days in the late fifties and early sixties. The cancers that were sure to follow from open-air testing—especially in Utah and Nevada and the supposedly empty, barely populated areas of the Great Basin of the American Southwest—only began to appear some fourteen to twenty years later. About 250,000 American GIs had been exposed to the tests during training, draftees forced to carry out maneuvers beneath the mushroom clouds. Similar numbers of American citizens, local residents of small isolated towns like St. George, Utah, where Hollywood filmed its Westerns, were also exposed. Groups like the National Atomic Veterans were formed, and groups like SANE continued their concern. The task, far more difficult, and still seemingly somewhat subversive, was to make the lives and deaths of the unfortunate families who knew firsthand the dangers of nuclear radiation tangible, visible, and viscerally important.[15]

Rachel Carson Is My Hero: Williams and Carson's Broader Ecological Vision

Though born in California in 1955, Terry Tempest Williams grew up as a child in Utah, downwind from the site of U.S. nuclear bomb tests. The explosions haunted her dreams, her fragmented earliest recollections. The abstractions of policy and national security were brought directly into the consciousness of this Mormon girl, who would eventually grow up to make such intrusions and their effects quite palpable—seen and felt through a combination of science and searing stories.

Terry Tempest Williams is a writer, naturalist, and activist. She is an award-winning author with a substantial body of work, a trained scientist who believes that science without feeling is never enough. In addition to the dangers of nuclear weapons testing, she has written about the western desert, the need for open democracy where the human heart has a say, on Hieronymus Bosch's heaven and hell, and other subjects. In each case, it is the need for observation, openness, listening, imagination, and empathy that are central to her work. Like Rachel Carson's, her vision is interdisciplinary and ecological, combining a love of birds and nature, science and society, wilderness and words, in an evocative, poetic, almost spiritual prose. Like Rachel Carson, she is also shrewdly political, working with environmental, antinuclear, and health groups, talking and testifying to the Congress when her voice can make a difference. And like Carson, she is not alone; she writes and acts as part of a movement; she descends from a line of women that includes Rachel Carson and many who came before her long ago.

In her 2002 essay, "The Moral Courage of Rachel Carson," marking the fortieth anniversary of *Silent Spring*, Terry Tempest Williams puts it simply. "Rachel Carson is a hero of mine." At about age seven or eight, she recalls, it was from Mimi, her beloved grandmother, Kathryn Blackett Tempest, that she first heard mention of Carson. "We were feeding the birds—song sparrows, goldfinches, and towhees—in my grandparents' yard in Salt Lake City, Utah. 'Imagine a world without birds,' my grandmother said as she scattered seed and filled the feeders. 'Imagine waking up to no bird-song.' I couldn't. 'Rachel Carson,' I remember her saying."[16] Williams later discovered a copy of *Silent Spring* and read it decades after her childhood memory. She felt kinship and shared moral indignation immediately. "Her voice is graceful and dignified, but sentence by sentence she delivers right-hand blows and counter-punches to the status quo ruled by chemical companies within the kingdom of agriculture." Williams writes of Carson, "I want to carry a sense of indignation inside to shatter the complacency that has seeped into our society. Call it sacred rage, a rage that is grounded in the knowledge that all life is intertwined. I want to know the grace of wild things that sustain hope."[17] She asks if we can find the moral courage that Rachel Carson showed and sees it as "this American tradition of bearing witness." She quotes Carson approvingly from one of her many principled and patriotic-sounding passages in *Silent Spring*, "If the Bill of Rights contains no guarantee that a citizen shall be secure against lethal poisons distributed either by private individuals or public officials, it is surely only because our forefathers, despite their considerable wisdom and foresight, could conceive of no such problem."[18] Despite such views deeply rooted in American tradition, Williams reminds us, Carson was called a spinster and a communist, and she was attacked by industry and by the chemical and medical establishments. But she fought back, spoke out, held her ground, showed moral courage.

But Terry Tempest Williams is not all indignation; there is inspiration, too. Like Rachel Carson and her predecessors in Victorian times, Williams softens moralism with imagination, with a clear intention to reach wider, more popular audiences and to bring them closer to something that once

FIGURE 4.1   Terry Tempest Williams, "Rachel
Carson is my hero." Photograph by Mark
Babushkin.

was called the holy. "Rachel Carson's name is synonymous with courage. . . .
But perhaps Carson's true courage lies in her willingness to align science
with the sacred, to admit that her bond toward nature is a spiritual one."[19]
Carson and women writers before her have been accused of sentimentality, naïveté, soft science, anthropomorphism in describing animals, and
worse. For Williams these are accolades, not assaults. Again, she quotes her
"hero" with admiration: "I am not afraid of being thought a sentimental-
ist when I say I believe natural beauty has a necessary place in the spiritual
development of any individual or any society. I believe that whenever we
destroy beauty, or whenever we substitute something manmade and artificial for a natural feature of the earth, we have retarded some part of man's
spiritual growth."[20]

Rachel Carson and Other Foremothers

Terry Tempest Williams is not only a metaphorical or spiritual daughter of Rachel Carson. She has literally walked in her footsteps and those of earlier women. Some of her earliest memories are of birding with her grandmother at the Bear River Migratory Bird Refuge in Salt Lake City. Rachel Carson actually visited Bear River in 1947, less than a decade before Williams was born, with her artist friend from work at the FWS, Kay Howe. They arrived by train in Salt Lake City as the last stop on a western trip to do research and gather firsthand observations for the government educational pamphlet series Conservation in Action. Carson and Howe toured by airboat and found the water so shallow at Bear River that the vast numbers of crowded waterfowl had spread botulism. Rachel was furious as her guide would haphazardly and unscientifically occasionally reach out and inject something into a particularly sick-looking bird. It is why her government pamphlets combine lucid, lyrical writing with sharp observation and description, basic facts and science, and reminders of the importance of understanding the complete ecology of any area. The resulting booklet, *Bear River*, which Carson wrote with Vanez Wilson and was illustrated by Bob Hines, came out in 1950. Like all her others before her fame, it was used and reused for years and considered one of the best FWS publications ever.[21]

Linda Lear tells us that Carson researched each area and each National Wildlife Refuge she visited intensely and thoroughly. Though we cannot be certain, she likely was aware of the writings of her predecessors, Florence Merriam Bailey and Olive Thorne Miller, who also traveled by train to Salt Lake City and the Great Salt Lake in 1893 as part of an effort to restore Merriam's health with the clean, dry air of the mountain West. The two companions traveled, birded, and took notes together. Each published the results later, adding to the lure and appeal of the West and its splendors for their predominately East Coast readers. Merriam's *My Summer in a Mormon Village* focuses mainly on her observations of life in Utah, although observations of nature and her concern to save it are sprinkled throughout. Her observations on killing for sport are typical:

> We met a party of summer hotel young men, calling for a gun—they saw grouse in the trees. I heard them recalling their recent achievement—they had killed a badger, a deer, and an eagle within a few days. It was a rude shock to

me, and I thought bitterly that even these wild grand mountains would soon be "civilized" by the pleasure-seekers who destroy all they can of the nature they have come to enjoy. . . . I could only reflect thankfully that though the mountains might be made patent-medicine advertisers, and the deer that drank from the lakes at their feet and the eagles that soared over their heads might be killed to gratify man's lust of power, the cloudless blue sky above us was beyond their reach.[22]

Of course the sky would not remain out of reach; the airplane, aerial spraying, widespread air pollution, and global climate change were yet to come. But Florence Merriam Bailey's viewpoint is clearly ecological; it goes far beyond even the activist concerns that she and Olive Thorne Miller demonstrated in trying to halt the millinery trade's slaughter of herons, egrets, and other birds. Mrs. Miller returned east to write more specifically about birds and nature in Utah in her popular 1894 book, *A Bird-Lover in the West.* She offers an entire section called "Beside the Great Salt Lake" that charmingly takes us on bird walks with her friend (Florence Merriam Bailey) in search of the nests of magpies and chats, while marveling over Western Bluebirds, Tanagers, Lazuli Buntings, and more. But it is her feelings about the gulls at the Great Salt Lake and her admiration for the Mormon view that they are sacred that foreshadows Terry Tempest Williams.

Miller writes about the omnipresence and tameness of the Ring-billed Gulls she sees following the Mormon farmers in the fields and how "one is tempted to believe the solemn pronouncement of the Salt Lake prophet, that the Lord sent them to his chosen people." The occasion for this belief among the Latter Day Saints, which is also reported in Williams's *Refuge,* is, according to Miller,

the advent, about twenty years ago, of clouds of grasshoppers, before which the crops of the Western States and Territories were destroyed as by fire. It was then, in their hour of greatest need, when the food upon which depended a whole people was threatened, that these beautiful winged messengers appeared. . . . The crops were saved and all Deseret rejoiced. Was it any wonder that a people trained to regard the head of their church as the direct representative of the Highest should believe these to be really birds of God, and should accordingly cherish them? Well would it be for themselves if other Christian peoples were equally believing, and protected and cherished other winged messengers, sent just as truly to protect their crops.[23]

Miller goes on to note approvingly Joseph Smith's levy of a penalty of five dollars for any gull that is shot in the Territory, that they now nest secure on the solitary islands in the lake, and that "no man or boy dreams of lifting a finger against his best friend." This is extraordinary, she says, as she is used to seeing every other bird who renders a like service, "shot and snared and swept from the face of the earth."[24] These environmental regulations, if you will, based in a religious appreciation for the gulls, strengthen the respect and admiration Miller and Bailey have for the Church of the Latter Day Saints, among them the great-grandparents of Terry Tempest Williams.

Such mixtures of scientific observation, nature writing, travelogue, and spiritual and moral values are hard to find these days. They were, in fact, harshly criticized by male, increasingly specialized experts even then. William Brewster, the noted ornithologist, had, in fact, attacked Olive Thorne Miller for publishing her discovery of a new bird species in the popular literary magazine *Atlantic Monthly*. Brewster was probably embarrassed and angered because he announced his later finding of the same species in the ornithological journal the *Auk*, only to learn that Mrs. Miller had already written about it in the *Atlantic*; her scoop had also been repeated when her article was assembled, along with other pieces of her writing, into *Little Brothers of the Air*. He chastised her for scattering her findings in popular magazines and for burying them in books with titles like *Little Brothers*.[25] Replying in a letter to the *Auk* with ladylike restraint, Miller refers to Brewster's "gentle admonition" and then delivers a defense that is devastating while demure. It could stand today, along with Carson and Williams, as a paean to the importance of popular, poetic, and precise nature writing that is rooted in the delight of discovery, the importance of images and imagination, and an insistence that each living creature is a wondrous creation. Olive Thorne Miller explains why she writes in a literary, rather than a narrowly scientific fashion:

> There is first, my great desire to bring into the lives of others the delights to be found in the study of Nature, which necessitates the using of an unscientific publication, and a title that shall attract, even though it may "ambush" my subject. I have never studied scientific ornithology. . . . Let those who will spend

their days killing, dissecting and classifying; I choose rather to give my time to the study of life, and to doing my small best toward preserving the tribes of the air from the utter extinction with which they are threatened. And lastly, a confession: I should take pleasure in "sharing my discoveries" were I so happy as to make any; but to me everything is a discovery; each bird on first sight, is a new creation; his manners and habits are a revelation, as fresh and interesting to me as though they had never been observed before. How am I to tell what is an old story and what a new one? Study these things who will. I study the beautiful, the living, the individual bird, and to my scientific confreres I leave his skin, his bones, and his place in the Temple of Fame.[26]

Both Florence Merriam Bailey and Olive Thorne Miller also wrote broadly about nature in the West. Interestingly, Bailey had written an article for the *Condor* in 1907 on the White-throated Swifts at Capistrano, where Terry Tempest Williams was born and has her earliest recollections. Even late in her career in 1939, Florence Bailey published a widely used and often reprinted Department of the Interior pamphlet on the *Birds of the Grand Canyon*. Terry Tempest Williams also follows in a line of other western environmental women who sought broader audiences and influence. As a long-time naturalist educator at the Utah Museum of Natural History, she also descends from Martha Maxwell of Colorado and her lifelike presentations of animal specimens in ecological settings at her Boulder museum and at the 1876 Philadelphia Centennial.[27] Her political activism grows out of earlier efforts like those of the Utah Women's Club that organized to preserve nature, including birds, wildflowers, and more and advocated for parks and preserves in the early twentieth century. Williams herself has acknowledged the influence of environmental values from frontier Mormon women.[28]

But of western writers, Terry Tempest Williams is perhaps most like Mary Austin, for whose *Land of Little Rain* she has written an introduction. Williams writes of her, "The testimony to Austin's brilliance as a writer is in the timelessness of her prose. 'This is the nature of the country. . . . There are hills, rounded, blunt, burned, aspiring to the snowline.'"[29] For Williams, foremothers like Austin are constantly present. "Mary Austin haunts me. I am intimidated by her. She is a presence in my life even though she has been dead for sixty years. I have a photograph of her that sits on my desk." She explains how Mary Austin wrote to the famous nature photographer, Ansel Adams, who shot a portrait photo of her, not to thank him, but to complain in some detail what she did not like about it. As Williams says, "This image of Mary Austin makes me smile. She was tough, cantankerous, and hardly gracious. . . . I love that Mary Austin was not polite or coy or particularly accommodating. Too many women have been silenced in the name of 'niceness.'" Williams admires Austin's spiritual nature, her love of the desert, and her combination of evocation and environmental advocacy. "In many ways, I view her as a sister, soulmate, and literary mentor, a woman who inspires us toward direct engagement with the land in life as well as on the page. She was unafraid of political action embracing the rights of Indian people, women, and wildlands. Mary Austin was a poet, a pioneer, and a patriot."[30]

Terry Tempest Williams blends the influences of Mary Austin, Rachel Carson, and others into a unique modern style that combines scientific talent and training, strong moral norms, and the capacity for outrage, alongside a strong sense of storytelling, sensitivity, wonder, imagination, and a love for nature. Like Rachel Carson, these qualities give her appeal to audiences far wider than those of natural scientists or politicized environmentalists. Carson was raised in a firmly Presbyterian household with deep roots in Protestantism, including ministers; Williams has grown up in a leading Utah Mormon family, with close ties to the church going back to the 1840s; her line includes male elders and other church officials. Over time, each of these two women redefines her religious roots into a more modern universalism, as nature takes on a sublime quality that may strike the more orthodox as transcendentalism, even paganism. Each woman draws most from older values instilled by mother, grandmother, and other female ancestors. This means that for all her ease with contemporary progressive values, Terry Tempest Williams is in many ways a true conservative, or at least the inheritor of nineteenth-century values of the importance of family, ancestry, place, and a religious approach to the connectedness of humans and the heavenly here on earth.

Terry Tempest Williams and Family

Terry Tempest Williams's values, like Rachel Carson's, were formed at an early age and are rooted in family history and the influence of her mother and grandmother. Terry went to the beach at Capistrano in California with both of them when she was very young. It is at the beach, with its vistas and crashing waves, that she learned about life on the edge of things, about transitions, and the simultaneous beauty and danger of the ocean and of nature. Looking back, she recounts, "And it is here, I must have fallen in love with water, recognizing its power and sublimity, where I learned to trust that what I love can kill me, knock me down, and threaten to drown me with an unexpected wave. If so, then it was also here where I came to know I can survive what hurts." It also created a special bond with her mother Diane Dixon Tempest, whose quietness and love for Terry gave her the space to flourish and speak out. "My mother and I came to trust each other on the beach where we sat. Between the silences, we played together. We entertained ourselves. On the edge of the continent, looking west, we came to an understanding of the peace and the violence around us."[31] But it is her grandmother, Kathryn Blackett Tempest, who links her to the nature-study movement, to a Mormon ethic that cares for the land, and to the world of spirit.

In her essay "The Architecture of a Soul," Terry Tempest Williams begins by intoning the names of shells, "Pink murex. *Melongena corona*. Cowry. Conch. Mussel. Left-sided whelk. Lightning whelk. True-heart cockle. Olivella. *Pribilof lora*. Angel wings." They are among those that Terry and Mimi collected and catalogued when she was eight years old. They pored over field guides checking their identifications. Now they still rest in a basket in Williams's study, an "inheritance" from her grandmother. "They remind me," she says, "of my natural history, that I was tutored by a woman who courted solitude and made pilgrimages to the edges of our continent in the name of her own pleasure, that beauty, awe, and curiosity were values illuminated in our own home. . . . I can hold *Melongena corona* to my ear and hear not only the ocean's voice, but the whisperings of my beloved teacher."[32] Mimi was interested in all of nature, but especially birds. She had a collection of Roger Tory Peterson's various field guides when Terry was little. Williams still has her grandmother's well-worn copy of *A Field Guide to Western Birds*. And she still has her own, given to her by Mimi when Terry was only five.

Both Olive Thorne Miller and Williams recorded a sighting of the colorful Western Tanager in their own books. Williams recalled spotting it at a friend's childhood birthday party. The host seemed miffed that Terry had wandered off from the festivities. But, when confronted, she asked to call her grandmother. Mimi soon arrived and had the entire party out happily admiring Terry's find, the bird with the red head, yellow body, and black wings that adorns the cover of Peterson's guide. It is the very first book that Williams remembers taking to bed, under the covers, and reading with a flashlight.[33] Terry spent many happy hours with her grandmother in the Bear River Refuge, a huge wetlands on the edge of Salt Lake City. And when Mimi feared she might lose her sight, she taught Terry and her brother Steve to identify each birdsong, using an early Audubon set of recordings on bird cards put into a device called an "audible auditron."[34] It is why, as Williams recalled later in an interview, Rachel Carson made such a powerful impression on her. "I remember as a child, my grandmother read to me *Silent Spring*. It was incomprehensible to me that there could be a world without birdsong. Rachel Carson was not only an influence on me but felt like family because of the regard my grandmother had for her."[35]

Like Carson, Terry Tempest Williams combined a love of nature, a sense of wonder, and a literary imagination at an early age. In addition to exploring and reading with the women in her family, Terry began to find her voice and strengthen her love of words and poetry in an unusual way. Her fourth grade teacher had told her that she had a lisp and needed speech therapy. It was deeply embarrassing, but perhaps a reason why she hated reading aloud in class, ignored playmates' teasing, and often remained quiet. So, instead of hours at recess, she spent time with Mrs. Parkinson. In addition to tongue placement exercises, they read poetry aloud together, Terry's voice over hers. Emily Dickinson, Robert Frost, and, because her teacher knew that Terry loved birds, poems that feature them. "She taught me how to hear the sounds of words and find delight in the rhythm and musicality of certain combinations." For homework, Terry also read poems aloud with her mother, practicing elocution, speaking clearly. "Suddenly I began to enjoy the art of speaking, because it followed the art of listening. These poems were puzzles and secrets, each with its own hidden meaning. . . . My task was to honor the power of each word by delivering it as beautifully as I could." This is reminiscent of Rachel Carson's lifelong practice of reading her prose aloud, often to her mother, as she revised and sought the power and beauty of each word, each sentence.[36]

Like Rachel Carson, Williams loves both English and environmental science. As she puts it, "I guess I've always had a love affair with language and with landscape." She holds a BA in English, with a biology minor, from the University of Utah, along with a master's degree in environmental education. In college in the 1970s, she was torn because "there wasn't enough time to take all the English requirements *and* all the biological requirements." She asked her adviser if she could major in environmental English and he replied, "Absolutely not!" She asked the biology department if she could major in "literary biology," and they told her she was "completely mad." Only in graduate school was she able to integrate her interests through a study of Navajo culture that combines nature and narrative.[37]

Williams tested her theories in her first teaching job at the Carden School of Salt Lake City, a conservative private school founded and run by the prim and proper Mr. and Mrs. Jeff, members of the right-wing John Birch Society. There were dress codes, strong discipline. The word *biology* was not to be mentioned because it can refer to sex. *Science* was the preferred term. When she brought bones and feathers and other wonders from the natural world into class, she found them cleaned up and discarded by the Jeffs, who preferred neatness. There were a number of struggles and remonstrances in Mrs. Jeff's headmistress office. But Williams persevered, loved the children, and hoped to spark their creativity much as did her grandmother, or Anna Botsford Comstock and several generations of women in the nature-study movement. But when Mrs. Jeff came into Williams's classroom one day and to find her pupils rolling on the floor and pretending to be humpback whales, as a record of whale sounds played in the background, it was too much. Mr. and Mrs. Jeff demanded to know if Williams was "an environmentalist." When she admitted that she was, Mrs. Jeff exclaimed, "We had our suspicions when you and Mr. Williams went to Alaska and did not carry a gun." And Mr. Jeff added, "Did you know that the Devil is an environmentalist?"

Williams was fired, but when the Jeffs reflected about how much the children loved her classes, they relented—on the condition that she would never let the children know she was an environmentalist. Over time, however, Williams developed respect for Mrs. Jeff's ability to hold her classes spellbound with literature read aloud, asking the children to embellish the plots, or discussing leadership while reading Julius Caesar. And she also discovered more of her own voice and the power of following one's passion; she continued to have her children draw and write about the natural world and venture outside the classroom to observe and write about nature firsthand; she brought their finds back to the classroom so they could watch a praying mantis devour its mate and spin its egg sac that hatches in the spring. It was for her a direct connection to Rachel Carson, whom she quotes, "If a child is to keep alive his inborn sense of wonder, he needs the companionship of at least one adult who can share it, rediscovering with him the joy, excitement, and mystery of the world we live in."[38]

This is why Terry Tempest Williams's first book, like Carson's *The Sense of Wonder*, is for children. *Between Cattails* draws on her own childhood with her grandmother in the Bear River Refuge. It is, she says, "a celebration of my obsession with the Bear River Migratory Bird Refuge, dedicated to my grandmother, Mimi. It was the first place where I fell in love with birds. Again, I spent a year studying about marsh ecology since I wanted the text to be factually accurate. I also wanted it to sing like the marsh sings, so it was another exercise in figuring how to braid these different voices—the scientific and the lyrical."[39]

All these strands came together for Terry Tempest Williams—the wonder, science and story, the narratives of Navajo culture—when she learned from her father that her nightmarish recollections of atomic mushroom clouds had actually occurred. And that her mother's breast cancer, and that of her grandmother and other relatives, was related to repeated radiation exposures from the tests. It was a moment when she questioned many of her beliefs and the world of external authority. She identified with the wily coyote of Navajo legend and wrote *Coyote's Canyon,* a photo collection of the red rock Utah desert with her evocative text that identifies those who love and try to protect the desert as a secretive, subversive band. And then she wrote an essay called "The Clan of the One-Breasted Women," which combines her political concerns with her powerful family history. This piece became the basis for *Refuge.* "I know the things that concern me, and the things that concern me are relationships. Women, health, and the environment—no separation. So that the essay 'The Clan of the One-Breasted Women' was a seminal moment for me.... I had been watching the women in my family die, one by one ... and then to tell my father over dinner about that dream and having him say, 'You did see it.' 'Saw what?' 'The flash. The bomb. I thought you knew that. It was a common occurrence in the '50s.'" Terry's father thought a tanker had blown up. And then radioactive ash began to land on the car. "I think it was at that moment—talk about the power of story—that I realized the deceit I'd been living under.... The fact that the land, the health of the land, the health of the women in the land are all related ... that is how I approach environmental issues. It's

FIGURE 4.2   Terry Tempest Williams in Arches National Park, near Moab, Utah. Cheryl Kaufman Photography.

not an abstraction. Wilderness is not an abstraction, it's not an idea, it's a place."[40]

In the spring of 1983, almost twenty years exactly after the five nuclear tests in July 1962 that Rachel Carson warned about, Terry Tempest Williams began to record her observations and feelings about the effects on birds and ecosystems caused by record flooding of the Great Salt Lake and its Bear River National Wildlife Refuge, along with her family's simultaneous struggles with cancer. When she completed her journal, and after much agonized rewriting and editing on such sensitive subjects, the result became *Refuge*, her first major book. It resembles Rachel Carson in its love of birds, its blending of imagination, wonder, and poetic prose, and its sharp scientific observation, careful documentation, and principled pragmatism. But unlike the more reticent, private, and sedate Rachel Carson, Terry Tempest Williams, born September 8, 1955, came to adulthood in the 1970s, a product of Earth Day, the peace and women's movements, and the cultural shifts for women that followed. She reveals a startling openness and honesty about

her feelings and her personal life—whether discussing cancer or copula-
tion. But like Rachel Carson, who, on the surface, seems prim and proper by
comparison, more guarded about her intimate friends, her cancer, and her
political associations, Williams is far more contradictory and complex than
her exterior portrait might suggest.

As with Rachel Carson's tutelage by her mother, Maria, Terry Tempest
Williams was brought up with a strong matriarchal line. Raised in Mormon Utah, rather than Presbyterian Pennsylvania, she has been steeped
in nature, history, beauty, imagination and spiritual values from her earliest days. Her grandmother, Kathryn Blackett Tempest, to whom *Refuge* is
dedicated, doted on her and was determined to introduce her to the glories
and mysteries of the natural world, especially birds. Typical of Williams's
approach is the opening of *Refuge.* Here, within a few pages, we learn that
the author is marvelously forthright—discussing how she is the sole surviving woman in her immediate family, the rest having died of cancer. We find
that she knows history and science—as she explains the evolution of the
Great Salt Lake and its various rock formations and flood levels, and notes
its description by Captain Howard Stansbury in *Exploration and Survey of
the Great Salt Lake* (1852). We also discover that she is a feminist, though
belying stereotypes, an attractive, humorous, and irreverent one.

We are drawn in, even amused, as well as horrified, when we quickly
join Williams in a conversation about female rage with a close friend. They
are seeking out a site that harbors burrowing owls. The two women are
soon leeringly approached, hit upon, and condescended to by gross men
in a pickup truck who have obviously just destroyed the owls' nests. Williams gives them the finger and drives off; then, in a spirited retelling of the
encounter, she appalls her mother with her behavior. We soon find out,
too, that Williams is a very unorthodox Mormon woman who is close to
her wise, spiritual grandmother. She often dreams about or recalls her past,
her emotions, and the future. She is in touch with far more than most of
us. She recalls that as a child, "The days I loved most were the days at Bear
River. The Bird Refuge was a sanctuary for my grandmother and me. . . .
I recall one bird in particular. It wore a feathered robe of cinnamon, white,
and black. Its body rested on long, thin, legs. Blue legs. On the edge of the
marsh, it gracefully lowered its head and began sweeping the water side to
side with its delicate, upturned bill. 'Plee-ek! Plee-ek! Plee-ek!' Three more
landed. My grandmother placed her hand on my shoulder and whispered,

'avocets,' I was nine years old."[41] The experience is etched in consciousness, almost mystically, because, as Williams writes, "I was raised to believe in a spirit world, that life exists before the earth and will continue to exist afterward, that each human being, bird, bulrush, along with all other life forms had a spirit life before it came to dwell physically on earth. Each occupied an assigned sphere of influence, each has a place and a purpose. . . . We learned at an early age that God can be found wherever you are, especially outside. Family worship was not just relegated to Sunday in a chapel."[42]

Williams is able to blend her own apostate Mormon beliefs; her sensitive, almost magical perception; and her science in ways that appeal to the modern reader, even though they clearly are passed on from her grandmother, Mimi. Humor helps. Consider her precious, brilliantly recalled first trip to the Bear River Refuge with her grandmother, who treats her like an adult and a special friend. Seated side by side, Terry and Mimi take the Greyhound bus north to the refuge on US 91 with the Wasatch Mountains on their right. A classic birder, "a gray-haired pony-tailed woman," passes out the official checklist of birds in the refuge and urges the amateur ornithologists aboard to take "copious notes and keep scrupulous records." At ten years old, Terry pipes up, "what do copious and scrupulous mean?" Her grandmother says it means "pay attention." Terry busies herself drawing pictures of the backs of passengers' heads until an elderly man with a worn golf cap announces that in ten miles they will be "entering the Bear River Migratory Bird Refuge, America's first waterfowl sanctuary, established by a special act of Congress on April 23, 1928." Our established author, thinking back, says, "I was confused. I thought the marsh had been created in the spirit world first and on earth second. I never made the connection that God and Congress were in cahoots. Mimi said she would explain the situation later."[43]

When the Audubon guides on the bus to Bear River spot a group of White-faced Ibises, they call out in the fashion of birders intent on adding to their lists, "Ibises at two o'clock!" The bus jams to a halt and the bird club members pour out to look. To ten-year-old Terry, the ibises are a fresh discovery, a "revelation" in the words of Olive Thorne Miller. "Their feathers on first glance were chestnut, but with the slightest turn they flashed iridescences of pink, purple, and green." As more flocks arrive, "They coasted in diagonal lines with their heads and necks extended, their long legs trailing behind them, seeming to fall forward on hinges the second before they touched ground."[44] They probe the ground for food, with "their decurved bills like scythes disappearing behind the grasses. I watched the wind turn each feather as the birds turned the soil." This artistic observation by a child is interrupted by the no-nonsense didacticism of the expert guide. "Our leader told us they were eating earthworms and insects." In contrast, Terry's Mormon grandmother, Mimi, sees a spiritual, sacred scene. She "whispered to me how ibises are the companions of gods: 'Ibis escorts Thoth, the Egyptian god of wisdom and magic, who is the guardian of the Moon Gates in heaven. . . . When an ibis tucks its head underwing to sleep, it resembles a heart. The ibis knows empathy,' my grandmother said. 'Remember that, alongside the fact it eats worms.'" Terry's grandmother goes on to explain that Egyptians used the stride of the ibis as a measurement to build the great temples of the Nile; that the sight of them has pleased her; and that if need be, she could go home satisfied already. "The ibis makes the day." As they get back on the bus, Terry writes a succinct summary in her childhood notebook that anticipates her adult work; it combines facts with the fabulous, "one hundred white-faced glossy ibises— companions of the gods."[45]

As Williams muses and records her feelings and observations throughout *Refuge*, the Great Salt Lake slowly floods, squeezing out her refuge and release. Each chapter is noted by the most important birds she has seen as she observes the changes in the ecology of the lake and by an exact measurement of its rise. For Williams, "the refuge has been a constant. It is a landscape so familiar to me, there have been times I have felt a species long before I saw it." Or put another way, "The birds and I share a natural history. It is a matter of rootedness, of living inside a place for so long that the mind and imagination fuse." She has felt the magic of the place and counted on the familiar, yet always fresh, return of the ibises, Long-billed Curlews, the occasional whimbrel. Like Olive Thorne Miller before her, each sighting is an appreciation of a new creation. Terry Tempest Williams is modern and scientific; she measures the waters' flows and heights, the kinds and numbers of species seen. But she sees more and feels far more deeply than that. We soon discover that the rise of the Great Salt Lake and the flooding of the marshes and dikes are also parallel to and a metaphor for her mother's growing cancer. Neither has been anticipated, nothing can really prepare for such unwelcome changes. Diane Williams has been in remission from breast cancer for nearly a dozen years. Her daughter says of the lake, "I could not have anticipated its rise." This is quickly followed by, "My Mother was aware of a rise on the left side of her abdomen."[46]

Williams is wrestling here with change, unwanted change. The goal for our time is to find a way to feel, live, and act in a world where the old constants of beauty and nature and family can be broken unexpectedly, unnaturally. "It's strange to feel change coming. It's easy to ignore. An underlying restlessness seems to accompany it like birds flocking before a storm. We go about our business with the usual alacrity, while in the pit of our stomach there is a sense of something tenuous."[47]

As Williams recounts the natural history of her beloved Utah, we learn about ancient Lake Bonneville, pioneer explorers, the coming of the Mormons, the rise of cities, science, and of what we call civilization. We know what birds have been seen, noted, counted at every stage before. Williams recounts the checklists, the occurrences of the birds at Bear River Migratory Bird Refuge over the years, and how it is done. Those birds that appear rarely, randomly, are called accidentals. They show up, according to William H. Behle, author of *Utah Birds*, perhaps every fifty years, or once or twice since 1920. There must be recorded proof. Of such sightings at Bear River, there have been three American flamingos. These rare, accidental sightings contrast, for Terry Tempest Williams, with the artificial world we have created. "I personally have seen flamingos throughout the state of Utah perched proudly on lawns and in the gravel gardens of trailer courts. These flamingos, of course, are not *Phoenicopterus ruber*, but pink, plastic flamingos that can easily be purchased at any hardware store. It is curious that we need to create an environment foreign from our own. In 1985, over 450,000 plastic flamingos were purchased in the United States. And the number is rising."[48] Such manufactured birds are "our unnatural link to the natural world. The flocks of flamingos that Louis Agassiz Fuertes lovingly painted in the American tropics are no longer accessible to us. We have lost the imagination to place them in a dignified world." If a few wonders like a flamingo or a Roseate Spoonbill do appear, we consider them merely "accidental." The checklists and scientists carefully delineate even more imprecise, fleeting categories. There are species that are only "hypothetical" because there is not a provable, authenticated record from a competent observer. At Bear River, these include Reddish Egrets and Black Oystercatchers, Harlequin Ducks and scoters, Parakeet Auklets, and other marvels. Such scientific noting is necessary, but not sufficient. "How can hope be denied when there is always the possibility of an American flamingo or a roseate spoonbill floating down from the sky like pink rose petals? How can we rely solely on the statistical evidence and percentages that would shackle our lives when red-necked grebes, bar-tailed godwits, and wandering tattlers come into our country?"[49] For Terry Tempest Williams, quoting Emily Dickinson, "Hope is a thing with feathers that perches in the soul."[50]

Refuge offers reveries, revelations, and raw data throughout its evocation of our encounters with the cycles of life and death, of nature and nurture. There are almost unbearably poignant letters from Williams's mother to a new, young friend dying of cancer. Soon, grandmother Mimi is also diagnosed with terminal cancer. Each woman struggles before us, beside us, inside us, with the realities of dying. Death is natural and must be faced; what must *not* be accepted and what must change are the unnatural histories of family and place that are laid before us. Williams writes, "I am slowly, painfully discovering that my refuge is not found in my mother, my grandmother, or even the birds of Bear River. My refuge exists in my capacity to love. If I can learn to love death then I can begin to find refuge in change."[51]

It is only in the final chapter of *Refuge*, its "Epilogue: The Clan of One-Breasted Women" that we fully comprehend what we have been so slowly, sensitively prepared to experience. The Utah history, the paleogeology of the Great Basin, ancient native peoples, Mormon ancestors, wise women from great-grandmother to grandmothers to mother and on to grown-up child, flooding and loss of habitat, marvelous mysteries of birds, the foolish Salt Lake City plans to pump out and "save" the lake, the excruciating excursions into the cool, sterile cancer wards, agony and anger, pleas and prayers, the distant decisions of our own government that define seemingly blank areas on a map as suitable for bombing—all these transcend the prosaic piling up of proper facts. It becomes clear that Williams and her family are "downwinders." They have been showered with radioactive fallout over the years from the atomic test site southwest of them in nearby Nevada. More than two hundred of these open-air blasts were carried out between January 27, 1951, and July 11, 1962. The memories of mushroom clouds from Terry's childhood are the stuff of fragments and of dreams. But they are as real as are the deaths that now surround her more than twenty years later, which reflect the lapse of cancer latency periods of one or two decades after exposure to radioactive fallout.

Only at the end, when we truly know Terry Tempest Williams, when we feel and understand what is inside her, the history and heart of this writer, do we get her final story. It is of women going to the Nevada Test Site to bear witness, to be arrested. It is not a militant or mindless act of casually carrying out civil disobedience. It is an act of hope and honesty, of reclaiming the desert, of writing and revealing the stories of all the creatures that live and die there; it is rebuilding a refuge. As Williams is arrested, one officer handcuffs her wrists, and another frisks her body. "She found a pen and a pad of paper tucked inside my left boot. 'And these?' she asked sternly. 'Weapons,' I replied. Our eyes met. I smiled. She pulled the leg of my trousers back over my boot. 'Step forward, please,' she said, as she took my arm."[52]

Williams's Vision and a Reverence for Animal and Human Life

Terry Tempest Williams sees writing as the integration of her values and her vision, as an evocation of ecology, as a way to draw others, through accurate reporting and aesthetics, into a world of advocacy and of action. Like Rachel Carson, whose moral courage she admires, she relies on imagination, wonder, empathy, as well as science. And like Carson, she sees life whole. What sometimes seem to be scattered pieces, separate species, and specialized concerns form a glittering ecological and artistic mosaic in the eye of either an ecologist or an artist. This is the theme of Williams's *Finding Beauty in a Broken World* (2008). On the surface, *Finding Beauty* is simply reportage on three of her seemingly distant, disconnected research trips— to a mosaic studio in Italy, a prairie dog preserve in Utah, and wartorn, traumatized Rwanda. Described in terms of causes or concerns, the book is about three separate issues—art and aesthetics, animal rights, and human rights. Yet this somewhat unusual combination of concerns, akin to Albert Schweitzer's philosophy of a "reverence for life," was also shared by Rachel Carson, though not often perceived or presented in this way.

In *Finding Beauty*, the links between issues and stories are like tiles in a mosaic. Mosaic is at once an ancient artistic form and a modern term for integrated ecological landscapes or an ecosystem. The book opens with Williams learning about the production of mosaics in the classical style from a master artist in Ravenna, Italy. As her teacher, Luciana, explains, the art is in a harmony of tensions. Mosaics are made from small, chipped, broken and shaped pieces of tile or rock called *tesserae*. Even the creation of a single one takes patience, practice, skill, careful attention to detail, and an eye for the uniqueness of each stone. From the start, words in Williams's narratives resonate with the richness of metaphor. These broken pieces will become prairie dogs, people in Rwanda, perhaps ourselves. "Tesserae are irregular, rough, individualized, unique," says Luciana.[53]

The earliest mosaics celebrated animals and gods of nature; they are alive with antelope, dolphins, deer. By Christian times, Williams explains, mosaics communicate powerfully through image and story to preliterate viewers, who are simultaneously informed and inspired. "Mosaics became everyman's Bible; one didn't have to rely on words, only the shimmering stories told through images above, meant to overwhelm the viewer as one would be in the presence of God."[54] What is being laid out here is an updating, a transmutation, of Transcendentalism in an age of trauma. Another master mosaic craftsman, Marco de Luca, tells Terry, "Part of the nature of man is to recompose a unity that has been broken. In mosaic, I re-create an order out of shards." Or, as Williams says simply, "I believe in the beauty of all things broken."[55] From Luciana again, we understand that "mosaics are made out of community." This is perhaps the fine arts version of John Donne's "No man is an island, no man stands alone." Or, it lets Williams praise the insights from author and activist Paul Loeb that what each individual, each citizen, does really matters—personally and politically—if done as part of a network, a community. It is isolation that kills the spirit.[56] Madame Luc, another mosaic master in Ravenna, who seeks to create modern masterpieces within a changing, yet timeless tradition, offers an additional adage.: "Here. Hear. We create the future through a rearrangement of forms, what we have learned from the past."[57]

Themes emerge from such tesserae. We cannot escape the past, our history; we must reshape it from the broken, even violent pieces. For Williams, as for Carson, this is relentlessly researching where we have come from, where we have been. It is the stuff of ancestry and evolution, of parents and grandparents, or predecessors and paleoscience. We also live and work and feel in community—often without words, but with imagination. And each piece, each part of a society, or an ecosystem, is as essential as the next. The history and appreciation of mosaic art in Italy is essential and has prepared us for our sudden plunge into the paleobiology and contemporary plight of prairie dog populations.

The Plight and Parables of Prairie Dogs

Prairie dogs, as Williams, the natural history museum teacher tells us, are from the Pleistocene epoch and its ecological mosaic. American bison roamed and ruled across the West. Their huge, endless herds and hoofs trampled the land, compressing and compacting it. Prairie dogs would, with their tunnels, aerate the land, loosen the soil, so that they and others might flourish. Forming huge prairie dog towns or cities, these Paleolithic populations, at their peak, created underground areas as large as 250 by 100 miles, or 25,000 square miles—homes for some 400 million prairie dogs. Such reclaimed land and loosened soil became home on the prairie to a host of other species, to spectacular biodiversity far beyond the American bison. More than 200 animal species were recorded, including 140 that benefited directly from the prairie dogs, such as the black-footed ferret.[58] The black-footed ferret uses prairie dog holes, but is less social, more elusive. Recall that in the nineteenth century, Martha Maxwell had collected one and sent it to the Smithsonian as part of the cataloguing of species carried out by Hart Merriam and Spencer Fullerton Baird; it had been mentioned by Audubon, but the experts had doubted his observation and its existence. But, unlike the ferrets, the sociable prairie dogs were often spotted as they popped up, almost preened, before pioneer observers. As early as 1804, Lewis and Clark were so taken by the cute critters that they saw on the plains "in infinite numbers" that they caught one and shipped it back alive to Thomas Jefferson in the White House, where he kept it as a pet. The naturalist Ernest Thompson Seton estimated in the early 1900s that there were five billion prairie dogs in the United States.[59] But the past is prologue; these interesting images and facts are presented as preparation for what follows. Williams is about to unpack for us, to flesh out if you will, a simple, objective, though chilling science story in the contemporary *New York Times*. In the December 6, 1999, edition, Les Eldredge had written "A Field Guide to the Sixth Extinction." The title is a reference to scientists' fears that our modern loss of species, our loss of biodiversity, due to climate change, habitat loss, pollution, and other pressures on animal populations is rivaling the fifth great extinction some 65 million years ago, when the dinosaurs went extinct along with thousands of other species; among those left was a small group of tiny, minor animals called mammals. Freed from the domination of the larger reptiles, mammals grew and grew and then finally took over.[60]

These scientific, biological, evolutionary facts are akin to those that motivated Florence Bailey, Mabel Wright, and Olive Miller to try to save entire populations of endangered egrets; they chose to encourage the lore and love of birds, to write for popular audiences, for children and the young, to tell a tale, to adopt through a little anthropomorphism the point of view of their favorite feathered creatures. Rachel Carson, too, took reports of the ravaging of robins, the obliteration of osprey, and the dangers of DDT and other deadly dustings and made of them a fresh mosaic of science and of art. And so it is we learn from Terry Tempest Williams that the Utah prairie dog, one of five somewhat similar remaining prairie dog species in the United States, has been labeled as one of six of the most likely species in our country to soon face total extinction. It is but a single tesserae. But properly shaped and set among the shimmer and stories of others, in a larger mosaic, it can move us and take on meaning. As we are first introduced to the Utah prairie dog, Williams characteristically has a dream. In it, she sees a small figure standing before her. It is a prairie dog. It speaks. "I have a story to tell," it says. Once awake, she wonders what the prairie dogs, with their huge eyes, see and smell as they stand and observe a group of men coming to poison their homes with toxic smoke, to rid them from ground in Cedar City, Utah, that is destined to become a golf course. We have shifted from our ordinary human perspective to the prairie dog point of view:

> They hear the sound of a truck coming toward their town, the slamming of doors, the voices, the pressure of feet walking toward them. From inside the burrow, they see the well-worn sole of the boot, now the pointed toe of the boot, kicking out the entrance to their burrow, blue Levi's bending down, gloved hands flicking a lighter, the flame, the heat, then the hands shoving something burning inside the entrance, something is burning, they back up further down their tunnel, smoke curling as the boot is kicking dirt inside, closing their burrow, tamping the entrance shut . . . they cough, they wheeze, their eyes are burning, their lungs are tightening, they cannot breathe, they try to run, turn, nowhere to turn. . . . The toxic smoke is chasing them like a snake herding them toward an agonizing death of suffocation, strangulation, every organ in spasm, until they collapse onto each other's bodies, noses covered in blankets of familiar fur, families young and old, slowly, cruelly gassed to death.[61]

This small, modern slaughter of some few remaining Utah prairie dogs has its roots in our recent, if forgotten, historic past. We learn in *Finding Beauty* that World War I produced a huge need for beef to feed our dough-boys overseas. Federal lands were opened to cattle grazing at government expense. Fortunes were made. In the process of creating grazing land, predators and little animals that created potholes were perceived as a hindrance to national security. They were labeled varmints that must be vanquished. "Killing rodents became a patriotic act." The extermination of the prairie dog was begun. The U.S. Biological Survey was put in charge. Ironically, or perhaps appropriately, this is the division that was under the supervision of Hart Merriam, the leading naturalist, collector, and friend of Teddy Roosevelt; he was also the brother of Florence Merriam Bailey—who chose to observe birds in the field, in life, rather than, like her brother Hart, shoot them for examination and classification in a lab. Williams offers us accounts in some detail of these early operations, including the caption from an old photo by Fred Patz put out by the Biological Survey, with the helpful caption that "over 132,000 men working afoot and on horseback in cooperative campaigns distributed 1,610 tons of poisoned grain on more than 32,000,000 acres of 'infested' range and farm land during the year 1920. The resulting destruction of prairie dogs and ground squirrels effected a saving of $11,000,000."[62]

Prairie dogs in Utah are now nearly gone—reduced to some ten thousand in scattered preserves—and still subject, under certain restrictions, to removal or killing to improve the use of private property. Williams not only lets us feel what it is like for a family of them to be exterminated and explains that when removed to new sites, most prairie dogs do not survive; she soon joins an observation team of naturalists studying them. This key middle section of *Finding Beauty* is introduced with epigraphs, including one from the Navaho Indians, who, when told of a plan to exterminate prairie dogs on their land, said, "If you kill all the prairie dogs, there will be no one to cry for the rain." The amused officials, we learn, assured the Navajo that there is no connection between prairie dogs and rain and killed them anyway. They were shocked at the outcome. "The desert near Chilchinbito, Arizona, became a virtual wasteland. Without the ground-turning process of the burrowing animals, the soil became solidly packed, unable to accept rain. Hard pan. The result: fierce runoff whenever it rained. What little vegetation remained was carried away by flash floods and a legacy of erosion."[63]

Williams also lets us know that she is part of a coalition of environmental groups, including the Southern Utah Wilderness Alliance and the Center for Native Ecosystems, petitioning to change the status of the prairie dog from "threatened" to "endangered." In order to do this, she believes more than petitions to the federal government will be needed. She writes in the *New York Times* about why we should be concerned about the fate of a furry rodent as war is about to unfold in Iraq, or compared with larger issues like terrorism, a shaky economy, racism, and the rest. Summed up, the answer is "Quite simply, because the story of the Utah prairie dog is the story of the range of our compassion. If we can extend our idea of community to include the lowliest of creatures, call them 'the untouchables,' then we will indeed be closer to a path of peace and tolerance."[64]

Barred from Bryce Canyon National Park because she has signed on to a lawsuit with environmental groups and because of her opinions published in the *Times*, Williams then joins a research project through which she is able to enter Bryce Canyon and study the interactions of prairie dogs up close. Along with three graduate students, she begins a stint with John Hoogland, author of *The Black-Tailed Prairie Dog*, as an assistant in his life's work of endlessly, relentlessly, scientifically observing prairie dogs. In his letter of invitation, Hoogland warns that she must be dedicated and prepared for discomfort. They sit quietly, endlessly, crammed into small plywood towers recording each prairie dog's movement and interaction. "While working with the prairie dogs, plan to work every day, seven days a week—at least during critical periods such as breeding and marking of babies. At less critical times I will make every effort to give you an occasional day off."[65]

What ensues is a bit like Jane Goodall among chimpanzees or Dian Fossey with the silver-backed gorillas, but attempted by an amateur. Williams records each failure, each frustration as a neophyte observer under these conditions. "Shaking.... I was useless.... He sent me to the far tower, which I had to climb. I am certain I will fall. Had to try twice. Finally made it with my pack on—felt like a turtle climbing a twenty-foot tree.... The box that will be my home for fourteen days measures four feet by six feet with large cut out windows." Her job is to never leave this plywood cage (there is a small hole for peeing, no need to climb down) and observe prairie dogs #35, #24, #R31, HWA, #70, and #RR6 among the 125 individuals who live here, one of the last "protected prairie dog populations in the world."[66]

The prairie dog preserve is near a work shed, trucks, and other human hullabaloo not far away. It is a broken tile on a lonely, lifted landscape some 8,100 feet above sea level on the Paunsaugunt Plateau. It is cold, and Williams is alternately shivering, cramped, bored, yet slowly fascinated by the village life that emerges beneath her. She watches all the prairie dogs, or p-dogs as they are called, simultaneously face the sun, as if in worship. Another feels the pulse of the earth with "his right paw on the ground, his left paw is open toward the sky." She hears and learns alarm calls with all the p-dogs on high alert, "clicking, chirping, all standing upright in the same direction with their backs arched, crying." She sees, feels nothing until more than two minutes later, a pronghorn antelope passes by.[67] When the winds howl and shake the tower, there are no prairie dogs to be seen. They are underground. The p-dogs match the exact clay color of the ground, yet the more Williams observes through her binoculars, the more colors emerge. She spots a shock of lapis on the ground. It is a Mountain Bluebird. There are robins, ancient bristlecone pines, and from her perch, she can see the edge of the Pink Cliffs. "It looks like a cut in the flesh of the forest. . . . Below, an erosional fairyland of sandstone spires appears. A thousand candles are lit in pastel shades of pink, orange and yellow."[68]

Soon she is helping to mark the p-dogs, noting that some are feisty, some are docile. She admits, unscientifically, to Hoogland that it is hard not to see them as "cute." She notes their eyes, the black nipples of the nursing females, the unique characteristics of each, like HWA, who has two distinct black stripes widely spread across her head. Hence her designation "Head Wide Apart (HWA)." It is this particular prairie dog who will especially win Williams's heart, but not before she writes sensitively of yet another cancer in the family; her brother Steve has been diagnosed with lymphoma. He has meditated and experienced peace by walking the labyrinth at Commonweal, a cancer retreat center near Point Reyes National Seashore. Can we extend such compassion and peace? Williams watches the p-dogs kiss, eat flax until the purple petals are strewn across the colony, notes that there is "something very tender about HWA." HWA begins to look at her. Terry focuses on her eye, which appears black, but "in truth, if you look at it long enough when the light is just right, prairie dogs have brown eyes, a deep amber color with a black iris. The eyes are shaped like pumpkin seeds. Head Wide Apart blinks. Her focus is straight ahead. She has a black eyebrow and below her eye is a faint streak resembling the dark stain of tears."[69]

Over time, Williams watches HWA try to help others who are being trapped to be marked for research, issuing alarm calls, guarding mounds. She is touched deeply by Head Wide Apart. "She is a small, lactating female who is calm, yet attentive. Her young have just emerged today. And at night, before she retires, she stands at the tower and looks up, before she focuses on the setting sun."[70] She watches HWA widen her burrow NX, building a protective mound around it, "like a fortress," and observes the litter of four pups tumble and play, scurry away from juncos and robins—making of it a game. These pups will grow to become individuals, just like those others that Williams says have "striking personalities and traits within the North Clan. P Dog #35 keeps everyone together, is always kissing; Madame Head Wide Apart is very tender, very sweet, eccentric in her behaviors, like always coming to stand beneath the tower and look up; #RR6 is an explorer, wanders far and wide into the woods and seems the most independent of the group; P Dog # 24 is somewhat of a loner and is most connected to P Dog #35 and Unmarked P Dog; Unmarked P Dog is elusive, refuses to be caught, cannot be identified with any one burrow or repetitive behavior. A maverick."[71]

Williams has clearly crossed some invisible line where a set of indistinguishable, scurrying rodents have become companions. She even wonders what their point of view must be, getting down on the ground to stare up at her tower. All of this is a wonderful, remarkable surprise to Williams— seasoned naturalist, sensitive soul, writer that she is. "I can't get #RR 6 out of my mind, how at dusk she sits on her haunches facing the setting sun with her palms pressed together—how can this not be seen as worshipful? Sun salutations. Many of the gestures we are observing are predictable and expected, but many of them are not. Surprise is part of each day. How else could I sit here largely rapt in a four-by-six plywood box on stilts and be mesmerized? I had no idea. I had no idea of the power of the prairie dogs, the force of their personalities and the impact that is theirs on this grassland-sage community on the edge of a ponderosa forest in the high plateaus of Bryce Canyon National Park."[72]

We find ourselves drawn in by the observations, the interactions, the community in this prairie dog town. Once they become known individuals, we cheer them on. One day a Red-tailed Hawk flies over and is mobbed by Stellar's Jays and robins. "Strong alarm calls from MADAME Head Wide Apart. Red-tail cries out, that unmistakable chilling call that echoes through the woods, as it perches in the old snag at the edge of the North Clan. . . . Hawk shrieks again. Flies. Soars over the Village. . . . Madame Head Wide Apart stands alone giving the alarm call, fearless, even as the hawk flies over her."[73] Williams writes of the consciousness of the p-dogs amongst whom she now lives. "To be able to witness the embodiment of a different kind of knowing, an intelligence that is not human, but prairie dog, is to realize we are just one consciousness among many. . . . The gift I have been given has been in the waiting and the natural passage of time. As a speed addict, it has taken me time to detoxify. But slowly, hour by hour, panic and boredom became awe and wonder. I grew quiet. I began to see, to hear, perhaps most importantly, I began to feel and believe that I could reconcile myself with another species by simply being present with them." Like Rachel Carson, she quotes Albert Schweitzer, "*I am life that wants to live in the midst of other life that wants to live.*"[74]

As Williams prepares to return home, she wonders:

> Will I see dear Madame Head Wide Apart before I leave? I want to thank her. . . . There she is—at the base of my tower—looking up. Patience from the Pleistocene. Instinctual knowledge is another kind of intelligence. Kind— another kind—*kind-ness*. What can I give her. I can give her my words. She looks up. On my hands and knees on the floor of my perch, leaning out the square cutout door, I meet her gaze. No words. I pack my belongings and quietly climb down the tower. She is still there, illuminated. I sit down on the sand, very near to her. Madame Head Wide Apart does not move or seem to mind my presence. She suddenly stands and faces the sun. We both do.[75]

This final parting of a person and a single prairie dog is remarkably moving once one has read each detail, each description of the diary set before us by Terry Tempest Williams. It is juxtaposed swiftly with the death and burial of her beloved brother Steve. There is no disrespect in the proximity, the placement of these fragments of her life. They reinforce, they reflect, these tesserae.

Williams's brother Steve had always been a collector of bones, of animal skulls. He showed them to his children so that they would be skilled at knowing and naming animals, but more important, at understanding and facing death as part of life, of finding continuity and hope even in a grave or crypt. And so we journey with his sister, an educator and curator at the Utah

Natural History Museum, to the vaults of the venerable American Museum of Natural History in New York. It is home to some of the earliest and most encyclopedic collections of animals in the world. As Williams tours and takes in this mausoleum of mammals, reptiles, and birds, the records and the bones almost come alive. We sense the importance of lovingly labeling and describing each specimen, of naming and remembering, of, in fact, honoring each individual. She has come to examine and learn more about the prairie dogs—now near extinction. The bones, the remains, are inscribed, polished, and preserved like works of art; the skulls and bones, the evolutionary adaptations of the prairie dogs are marvels in themselves; they are things of beauty.

The "mammal morgue," as Williams first describes the museum when walking in, is now transformed into a kind of Holocaust Museum. Amid the horror, the remains of the destruction of five billion prairie dogs, we can find some meaning, some hope if the story can be told, the narrative and names recounted. There is a resemblance of all animal species in death, as there is in life. Examining a mummified prairie dog, Williams remembers human mummies "housed in the museum at home," noting, "this prairie dog and the human beings I saw look eerily similar." Her specimen is a "Little Man," the name given to prairie dogs by the Navajo or Diné who live in the Four Corners of the American Southwest. She examines the jaw of a prairie dog from New Mexico and the Pablo Bonito ruins, which she has visited many times; she has even placed her hand "against the handprints left by the Ancient Ones." We are transported to a time, a place, a culture where there is no "other." She says simply, "The prairie dogs and the Pueblo peoples were no strangers to each other."[76] Here, amid the bones and skulls, we can feel her brother Steve's presence strongly; here, amid the prairie dogs and their story straight from the strictures of Deuteronomy. "I put before you life and death, a blessing and a curse. Therefore, choose life that you and your children may live."

Rwanda—A Broken World

"And so it was I entered the broken world," is the epigraph from Hart Crane that opens the final section of *Finding Beauty*. We are looking directly into the red-streaked eyes of a Rwandan woman who is leading Terry Tempest

Williams into a basement to a "pyramid-shaped glass case of bones rising from a floor of white square tiles. The bones: skulls, femurs, ribs, vertebrae are organized in rows, columns, piles."[77] Steve's spirit has taken us directly from the mausoleum of mammals, of prairie dogs in New York, to a carefully collected display of death, of human genocide in Rwanda. The pieces are falling into place. The deep, knowing, tear-stained eyes of this Rwandan woman, the tiles, the confrontation with death, the brokenness of life. "I didn't want to come. I didn't want to be in a place so familiar with Death," says Williams. "I had seen enough in my own family. I was also scared. The only thing I knew of Rwanda was genocide and the weight of that word. 1994, the year we Americans turned our backs."[78] But she has said yes to a Chinese American artist, Lily Yeh, who has created in the poorest part of Philadelphia a Village of Arts and Humanities. Yeh has "understood mosaic as taking that which was broken and creating something whole." She has taught community members to pick up shards of glass in littered lots and make art together, constructing a mosaic of a "A Tree of Life . . . on the standing wall of a building otherwise destroyed."[79] Now the Red Cross has asked her to design a genocide memorial in the village of Rugerero in Rwanda on the border of the Congo. And so Terry Tempest Williams works alongside Lily Yeh and her team for a full month in Africa. They begin with memorials in two churches and set out along red dirt roads, driving past wetlands where the Tutsi hid but were usually discovered—and then butchered by the Hutu militia known as the Interahamwe. In Nyamata, the church is stained with blood, full of holes from grenades, and filled with sacks of skulls. "Ten thousand people were murdered here."[80] Rwanda is difficult to accept and imagine. It is, says Williams, "a hell of our own making—those who killed and those who looked away. No surgical strikes, computerized or digitalized by military minds and top gun pilots, the eyes of these killers were on the eyes of those they killed. By hand. One million Tutsis were murdered in one hundred days."[81]

But as Williams observes and talks with individual survivors, hears their stories, the abstractions of the words—*Rwanda, genocide, Tutsi*—take on life and form and feeling. The history of colonization, the roots of tribal enmities, the ongoing slaughter across the border in Congo are no longer distant, devoid of life, designations of "the other." In this context, we meet a U.S. Marine, fresh from being wounded in Iraq. Upon recovery, he has been sent to Rwanda to protect the U.S. embassy because "the government doesn't like the vibes in Kigali." We are shocked, yet somehow sympathetic, as he asks Williams if she will help him out with something. "There was a war here, right?" He looks side to side. He is serious. "I mean, could you sort of fill me in on what happened?"[82]

Finally, the site for the genocide memorial is cleared, less than an hour away from Ruhengeri, the starting place of the trek to Volcanoes National Park where Dian Fossey and Amy Vedder did historic work with gorillas. Fossey believed that the park should not be preserved for the gorilla's sake alone. It had to be for the sake of the Rwandan people as well. Its mountains are where their water is stored, its preservation essential for farming and survival. *"American and European concepts of conservation, especially preservation of wildlife, are not relevant to African farmers already living above the carrying capacity of the land. . . . The farmers need to know, not too much about what foreigners think about gorillas, but rather that, 10 per cent of all rain that falls on Rwanda is caught by the Virungas and is slowly released to irrigate the crops below."*[83]

This combination of concern for wildlife, for water, for women who have been raped and killed, is the reverence for life espoused by Rachel Carson that led her to dedicate *Silent Spring* to Albert Schweitzer and to link the fate of the robins with the residents of towns across our land. It is why Florence Merriam Bailey chose to observe birds through an opera glass, to love them in life because what you learn to love you are less likely to kill. We must plumb our history, examine the bones, learn and imagine the names and faces, look deep into the eyes of those who have died, those whom we would kill. At last the genocide memorial is completed. A woman named Dorothee takes Williams into the Bone Chamber. "She has just finished painting the cement floor green and the walls turquoise. The coffins will be brought in tomorrow, covered with purple and white cloth, and placed on shelves. . . . When I am here, she says, I am not alone. I am here with my family." Others place the final stones in the mosaic, paint the beams; children sketch objects along with Rob, an artist and their teacher, as they learn words in Kinyarwanda and in English. Two street kids named Emmanuel and Innocent say, *"Umutima mwiza. . . .* The boys take his notebook and draw a heart and point to their chests. Friend/*inshuti*; love/*urukundo*; thank you/*murakoze*."[84]

Terry Tempest Williams is herself transformed as she finishes her tale. She has overcome many fears and found ways to fathom the deaths of so many in her family. There is a colorful procession on the annual day of mourning. Songs, tears, smiles, stories of survival. Her husband, Brooke, has joined her now. They have dedicated their lives to science, healing, writing, ecology. They have chosen, perhaps fearing for humanity's future, not to have children. After three decades of marriage, they have decided now to adopt a Rwandan child, Louis, descended from royalty long ago. His Tutsi parents, Michel Kinyungu and Kagoyire Annociata, want him to be safe, to have an education in America. They have worked and watched and stood with Terry. They have developed unspoken, instinctive trust in this woman who had known little of Rwanda, who had feared to come at all. Michel and Kagoyire have survived genocide, hiding, years of separation; Louis escaped disguised as a Congolese boy. Miraculously, they have been reunited. Now they want him to live and learn so that he can, understanding its past, return to Rwanda to lead the country to a fuller future.

Aesthetics. Animal rights. Human rights. Mosaics. Prairie dogs. Rwandans. All are linked, intertwined parts of a whole, of life itself. Rachel Carson's legacy is more than mere pesticide legislation. It is a love of life in all its forms and a call to understand, to feel, to imagine that each of us is a part of its mosaic, if we would but see. *Refuge* and *Finding Beauty* pick up where Carson's death from cancer had left us. It is as if Terry Tempest Williams, scientist, environmentalist, writer, humanitarian, has searched through *Silent Spring* and taken as her text Carson's warning that our abstractions, our unblinking desire for comfort and control, must change. She quotes Schweitzer once more, "Man can hardly even recognize the devils of his own creation."[85]

Ironically, Terry Tempest Williams has had her own courage, beliefs, and perhaps fears severely tested one more time. As I was finishing this book, she revealed in *When Women Were Birds*, her 2012 reflections on her own life, writing, and finding her voice, that she had suffered a stroke as a result of a kind of brain tumor called a hemangioma in the "eloquent" part of her brain. This section, Wernicke's brain, is the home of language comprehension and metaphor. Some of Williams's spiritual qualities and openness about life surely spring from her early awareness that, like many of the women in her family who have died of cancer, she has been exposed to significant levels of atomic radiation as a Utah "downwinder." She has reflected often on death and its meaning. But this is a surprise that she has never prepared for. In her book, she describes for her husband, Brooke, what has happened to her:

"the right side of my body was numb, my vision was blurred and my speech slurred." The surgeon tells her that her attack was caused by pressure on her brain from a small tangle of vessels; it can happen again. Lying in the hospital, she repeats, "This is not my story, this is not my story . . . What if I do have a brain tumor? How shall I live?" Williams is faced with an almost impossible choice between simply waiting to see what will happen to her in the future or undergoing risky brain surgery. After she asks the real meaning of his clinical description of the procedure, the surgeon says, "Meaning, we'll have to wait and see if you can understand what I say or can speak." She seeks second and third opinions. Each expert neurosurgeon asks the same question: "How well do you live with uncertainty?" Williams replies, "What else is there?" She opts to do nothing.[86] Like Rachel Carson, who kept writing *Silent Spring*, appearing on television, and testifying before Congress, even as she was dying of breast cancer, Williams chooses to continue, for her own sake, and for those who will come after her.

But if Williams differs from Rachel Carson in her emotional openness and the range of subjects she now pursues, she shares one more vital similarity. She steadfastly refuses to be seen in isolation, to even try to be "the little lady that started it all." She is proudly part of a broad ecological movement and, like Carson and her foremothers, to be pragmatically political in speaking truth to power. Gently set among the tesserae of her narrative, poetic pieces are glittering glimpses of her values, her activism, her confrontations with those who would destroy the earth and its inhabitants. Seen or set alone, such stones could be viewed as didactic, moralistic, clichéd. Their glare softened and muted by the musings that surround them, they are appealing in their forthrightness, even fortitude. It is no surprise, then, that like Carson, Williams has testified a number of times before Congress. She has also joined in marches, acts of respectful, nonviolent civil disobedience, and pulled together other artists in attempts to reach out to policymakers and the public. Linda Lear has collected some of Rachel Carson's writings of this sort—speeches, articles, testimony—in her collection called *Lost Woods*. With Terry Tempest Williams, we can see the same approach in her own small book, compiled with Stephen Trimble, called *Testimony: Writers of the West Speak on Behalf of Utah Wilderness* from Milkweed Press as part of its series Literature for a Land Ethic.[87] The volume includes poetry, short essays, and stories, all designed to create a Utah Wilderness Area of some 22 million acres, to save it from a reactionary bill that would split, shrink, and shrivel it to 1.2 million acres. For inspiration, the writers, assembled by Williams and Trimble, looked to a 1960 letter from Utah novelist Wallace Stegner to David Pesonen of the Outdoor Recreation Resources Commission about the wilderness idea, referring to the southern Utah lands of his boyhood as "the geography of hope."[88]

In the introduction, the twenty-one male and female writers say they are bearing witness, as did a group of writers and artists more than 120 years ago. That earlier group came to Washington to persuade Congress to designate the region of the upper Yellowstone as a national park. "They took scientific evidence, lyrical descriptions, and—above all—the stunningly powerful photographs of William Henry Jackson and the lush paintings of Thomas Moran, both of whom had been to the upper Yellowstone, with the government exploring expedition of Ferdinand Hayden, and combined them into a packet of information that they waved in the faces of presumably amazed and astounded senators and representatives until the lawmakers did the right thing: Yellowstone National Park."[89] Asked at their press conference what they would do if their efforts failed, Williams, Trimble, and the other artists replied, reflecting more than a century of tradition, "Writers never know the effects of their words. . . . We write as an act of faith."[90]

5 ⚡ THE ENVIRONMENT AROUND US AND INSIDE US

Ellen Swallow Richards, *Silent Spring*, and Sandra Steingraber

Hair pulled back into a tight bun, beneath thick brows, Ellen Swallow squints into the lens of the long, brass telescope. She uses her young, steel-blue eyes to observe the stars opening up before her gaze. This is the instrument, with its insights handed down from Galileo, that brought on a revolution in religion and in reading the evening sky. Swallow observes a small star cluster that even Maria Mitchell has missed. "Professor Mitchell!! There is something here we have not seen! Look, can you see it?" The aging Mitchell, mind undimmed, cannot perceive this novel point of light, but she swells with pride at her brilliant pupil's find. It is as if she has found it herself. The story will be told through the ages. The notes they make, the narrative they compile, will guide others in searches yet to come. Maria Mitchell has found a daughter as surely as if she had conceived.[1]

And so, Ellen Swallow Richards, upon the recommendation of the leading astronomer of her day, was admitted, conditionally, into MIT. But the instrument that she uses to open up new universes, to upend ideas and ancient beliefs, will not be a telescope in a tower. She considers following in the footsteps, filling the button-laced high-topped shoes, of Maria Mitchell. But she wants to be of immediate use to people, to serve on the earth beneath our feet, not in the ethereal, philosophical purview that has marked astronomy throughout the ages. Neither does she choose those other observational optics, the new, portable binoculars, pioneered by Zeiss and others, toted on trips to the field for ornithological observations by Florence Merriam Bailey. She will start her own revolution. For Ellen Richards, it will be her beloved microscope, given to her as a child, that opens new vistas, invisible in older times. She will be able, as if by magic, to peer at tiny, unknown organisms in the water, the air, the soil, inside the human body itself, the vital organs, the tissue, the bones. Ultimately, everyone in our time will be allowed to look and learn, to watch and wonder, at the changes, the chemistry inside of cells, the components of life itself.

Begun by Florence Merriam Bailey with *Birds through an Opera Glass*, the stories captured with binoculars, then by photography and film—all increasingly portable and powerful—spread amateur ornithology, the love of nature, the democratization of its science, far and wide. No need to shoot a specimen to collect it, nor trap it, as did Lewis and Clark, who sent a live prairie dog to Thomas Jefferson as a pet.[2] The desire to preserve and protect such marvels, rather than merely mount them in museums, was spread as well. By the early twentieth century, within the White House itself, Theodore Roosevelt, Lucy Maynard, Florence Merriam Bailey, and the members of the DC Audubon Club could watch on film the flight of birds and fear for their destruction. They could walk the trails along with concerned citizens to note the increase or decrease of birds and other species, to rise up to rescue them, to restore the places where they live. Rachel Carson and Terry Tempest Williams are in this line of descent from their foremother, Florence Merriam. They are naturalists, natural historians, birders, and field biologists, who bring us science and spin stories that tell us that human beings share the same fragile habitat as those we watch. We are linked together and with the land. It is binoculars, films, and photos that lead us from the intricate feathers and jeweled eyes of birds to the agony of their death; from the burrows, barking, and beauties of prairie dogs to their dreadful demise. Finally, we are led to stories of those we love and care for beneath the blasted skies of Utah, within the blighted, bloody ruins of Rwanda.

Rachel Carson in *Silent Spring* draws on the naturalist, conservationist, environmentalist tradition in which women have played a leading role. It is why we remember the robins and sometimes forget that Carson was equally concerned with human health. When put together, when binoculars are bound tightly—as in a chemical bond—to the miracle of microscopes, the environmental science called ecology fully emerges.

Historically, environmental science and environmental health grew out of other disciplines—epidemiology, cellular biology or histology, and chemistry. Epidemiology is the tracing of diseases to their source, often manmade in origin, through the mapping and statistical monitoring of cases in communities, cities, countries, even continents. Histology is the study of cells and how bacteria, viruses, toxins, and other minute marauders enter into our internal systems and make us sick. Chemistry identifies the structures and properties of compounds with which humans interact. In many standard histories of medicine, public health, and environmental science, the heroes are men. I still love and tell the classic account of Dr. John Snow, the Victorian British doctor who disproved the miasma theory of disease, which held that cholera came from dank, foul smelling vapors. Snow carefully traced and mapped each case to its actual origin—the water used by residents of East London who drew it from the Broad Street pump. He convinced city officials to shut it down. The cholera epidemic of 1853 was broken. But I like to add, that Snow was aided by a little-remembered Anglican pastor, the Reverend Henry Whitehead. He worked and served among the poor, was known and trusted by them. It was Whitehead whom mothers told in intricate detail where the diapers were washed, how their children sickened and died. It was thus that actual people, only then turned into dots of data upon a map, were saved by Snow and Whitehead together. Epidemiology and caring for the poor came first.[3] Then came confirmation, the discovery of the deadly germs themselves. Until the microscope, no one had been able to see the bacterium *Vibrio cholerae*, the cause of cholera, or the other microorganisms that bring contagion and disease. Epidemiology and cellular biology, together with chemistry, maps, and microscopes, would launch an environmental and public health revolution.

As these advances came, the germ theory of disease, the mapping of sources and of sickness, were first and best combined in the United States by Ellen Swallow Richards. She brought back from Germany and biologist Ernst Haeckel the idea and idiom of ecology. And with her desire to protect human life, instead of charting the heavens, she set out to test and to chart the spread of polluted water through the watersheds of Massachusetts. The maps she drew are still masterpieces of art and arcane knowledge made visible to all. A generation later, in the very early twentieth century, Dr. Alice Hamilton added the study of poor neighborhoods and workers, with their stinking industrial and urban chemical exposures, that made her Hull House neighbors sick. She combined sympathy, social work, and science to institute the origins of occupational and environmental medicine.

Her actions and advocacy, followed by Anna Baetjer, Harriet Hardy, Mary Amdur, and other foremothers, created toxicology, the modern study of poisons and how and where they make us ill. Before she chose her title, *Silent Spring*, echoing the poetry of Keats, it is why Rachel Carson called her research and writing on pesticides simply the "poison book."

In 1957, as Carson started work on *Silent Spring*, she drew deeply on developments in toxicology, epidemiology, and environmental health— cutting-edge science on radiation, chemicals, cancer. With her wondrous prose and prodigious imagination, she made clear how such things worked and why we, as well as the robins, were at risk. But her work was cut short by breast cancer of her own. In Carson's day, the mechanisms of carcinogenesis were less well understood, and one's personal cancer was rarely discussed in the light of day. The era of support groups and spirited marches was yet to come. The double helix had just been unraveled a few years before by Watson, Crick, and Franklin. We had not yet mapped the human genome, or looked deep inside of cancerous, crablike cells. But Rachel Carson antici- pated these things and drew on the best science available at the time. She would have welcomed and written about all that we have learned in our day about cancer and its environmental causes. She also would have embraced the ecological writings of Sandra Steingraber. Steingraber, a poet and biolo- gist, combines field observations and feelings, science and spirit, maps and microscopes, binoculars and biophilia, along with a talent for making sci- entific studies come alive through anecdote and astonishingly poetic prose.

Sandra Steingraber and Environmental Cancer

There are strong similarities between Rachel Carson and Sandra Steingra- ber. Both grew up in mostly rural, small-town America. Each watched her hometown horizons, in Pennsylvania and in Illinois, slowly stained and circled around by smokestacks. Both wanted to be writers from an early age, then blended their gift for sinuous sentences and solid science into acces- sible accounts for the average reader; Steingraber earned an MA in creative writing and published poetry before completing her Ph.D. in biology at the University of Michigan. Both Carson and Steingraber taught college biol- ogy for a time, before full-time careers as authors. Each produced a trilogy of books for which they are best known. Both women have brought alive contemporary, controversial concerns about chemicals, cancer, and the environmental causes of disease. Each has been actively political; each has served on environmental advisory boards within the government. Each has been part of a wider environmental movement, communicating with colleagues, citizens, and close friends around the nation, even the world. Each woman has faced cancer.

There are clear differences, too. Born in 1955, Sandra was a toddler when Carson began her poison book. Steingraber is a product of our time, the beneficiary of the women's, peace, and environmental movements. The personal is the political, and vice versa. Like her contemporary, Terry Tempest Williams, she is frighteningly forthright, whether about female fertility, family, friends, or the fear of facing death. She benefits, too, from the scientific advances and the broad interest in and access to environmental knowledge set off by *Silent Spring*. Sandra Steingraber wants to track down and help us learn about the possible environmental causes of her bladder cancer, developed as a youth. In the battle to prevent the poisoning of America, she is determined to grasp and raise the guidon that fell with the death of Carson. She is aware that she follows in a line that goes back to Ellen Richards and Alice Hamilton, to Carson and her friends like Shirley Briggs, who kept her work and legacy alive.

Steingraber's views are best summed up in a short, intensely personal and political essay called "*Silent Spring*: A Father-Daughter Dance," published in a collection edited by Peter Matthiessen in 2007 to mark Rachel Carson's one hundredth birthday. It opens with her longing to recapture the fondest memories of her father, who had died of Lewy body dementia. A World War II veteran who served in Italy and was heavily doused with DDT to kill the lice spreading typhus throughout the war-ravaged local population, Sandra's dad tried to put the war behind him when he returned to postwar Illinois. He taught consumer education and took up organic gardening— became a devotee of Rachel Carson and of *Silent Spring*, which he used in class. Sandra was a small girl at the time; she recalls dimly seeing an unusual edition of *Silent Spring*, a paperback with a green cover and line drawings. It was put out by the Consumer's Union, a special edition released to their membership when *Silent Spring* first appeared. Two years after his death, having finally read *Silent Spring* as an adult in graduate school, Steingraber is home and accidentally comes upon the old, green paperback, a talisman from her father. In her youth, dementia slowly robbed her father of his dignity; in the end, he was ranting and deranged. His death is for Steingraber not merely a cause for mourning; it is motivation and a metaphor. It is quite possible that her father's Lewy body dementia may have been brought on by chemical exposures. Various neurological and autoimmune diseases like Parkinson's and Alzheimer's have been linked to the use of pesticides. "For my father, *Silent Spring* was an antidote to wartime thinking. . . . The motto of his antitank unit—seek, strike, and destroy—might as well have been the advertising slogan for DDT and other pesticides that had been developed for wartime and then were aggressively marketed to farmers, housewives, and suburban homeowners after the war ended. Returning GIs were urged to grab a bottle of poison and go after dandelions, mosquitoes, and grubs. In demonizing the home front's new enemy, one ad even went so far as to place Adolf Hitler's head onto the body of a beetle."[4] Sandra's father finds the garden a respite from command and control. He loves puttering, nurturing, trying new things from gardening magazines. Organic gardening is sensible and scientific. So when Rachel Carson said things like "The 'control of nature' is a phrase conceived in arrogance, born of the Neanderthal age of biology and the convenience of man' her message resonated with my father. *Silent Spring* was his armistice."[5] In Boston, when Steingraber first starts out on her own poison book, *Living Downstream: An Ecologist Looks at Cancer and the Environment*, she joins a militant women's breast cancer march (before the popularity of pink ribbons) and has a placard reading "Rachel Carson Was Right!" thrust into her hands. It shows up prominently in a picture of her in the *Boston Herald*. She says that at that point her father would have been too deranged to recognize it. But the truth is, she says, "I am fearful of his disapproval. . . . A lifelong Republican, he did not associate the demure and dignified Carson with protest marches, for which he had little patience. . . . *Silent Spring* was my father's armistice. It was my call to arms."[6]

It is this combination of the personal and the political—Rachel Carson, a product of her times, kept them assiduously apart—that distinguishes Sandra Steingraber. She wants us to know that both she and her father may have suffered because pesticides and other poisons were hawked to homes in happy postwar America. We must go upstream from our homesteads with her to search out the social and scientific causes of soaring rates of cancer. And when we find them, it will take citizen engagement in politics in the broadest sense—from protest to pragmatic policy proposals—to prevent more people from being harmed.

It is this putting of herself, the reader, and Rachel Carson squarely in the middle of a movement that is important for Steingraber. Even though I had known and worked with Sandra before starting *Rachel Carson and Her Sisters*, I had not read "A Father-Daughter Dance." It states the central thesis for her work, as it does mine. She writes, "I have become fascinated by the evident reciprocity between environmental activism and *Silent Spring*. Carson was as illuminated by activism and advocacy as the contemporary environmental movement was influenced—some would say inaugurated—by the publication of *Silent Spring*. If I'm right about this, it means the popular portrait of Carson as a lone, impartial, above the fray genius requires some revision."[7]

Steingraber recalls for us that Rachel Carson thanked her friend and fellow birder Olga Owens Huckins in Boston for writing to her about the agonizing death of songbirds from aerial spraying. Their interactions have been well documented by Linda Lear. But Steingraber wants to highlight, to underscore, that Olga Huckins was not merely a concerned citizen, suddenly appalled. Steingraber shows that she was also deeply involved in an organized campaign. She was an organic gardener, editor, and member of the Committee against Mass Poisoning. She and its members wrote letters to many newspapers in New England and on Long Island, brought lawsuits, and were quite polemical, taking "a human rights approach to environmental harm." Olga Huckins condemned pesticide spraying as "inhumane, undemocratic, and probably unconstitutional."[8] But Steingraber's important insight here is that when the lawsuit by the Committee against Mass Poisoning reached the Supreme Court, it "became a magnet for media attention." It caught the eye of the well-known essayist E. B. White, who then interested *New Yorker* editor William Shawn. White suggested that Rachel Carson should do a piece, not him. When Rachel was offered a three-part, 50,000 word serialization, *Silent Spring* was born. Steingraber says, "In short, environmental activism in the 1950s opened up a critical space in the publishing industry for environmental writing, and that development, as much as the slow accumulation of scientific knowledge, was the genesis of *Silent Spring*."[9]

Sandra Steingraber also believes that her own writing, like Carson's, was made possible by openings pried apart by activists, as well as by the slow accretion of scientific knowledge. "The early 1990s redefined breast cancer awareness to include environmental awareness. Women at breast cancer rallies took to waving copies of *Silent Spring* as they marched. As a young biologist and also a cancer survivor I became interested." Prompted by an invitation to speak about *Silent Spring* and Carson's life as a cancer patient, Steingraber finally read and reread *Silent Spring*, corresponded with Linda Lear, began her trips to archives and Carson sites. Familiar with the growing scientific literature on links between toxic exposures and human health, Steingraber was also aware how little was known to the public. "I began to wonder if I could write books that, like *Silent Spring*, constructed a bridge over that breach. . . . I had been a biologist by day and a poet by night. But perhaps there was a way of bridging this breach as well."[10]

Sandra Steingraber and the Art of Tracing Cancer

Steingraber draws beautifully with words. Her book, *Living Downstream*, opens with the landscape of central Illinois stretched out before us, its vast sky like a planetarium filled with stars. She is proud to show this land to visitors who are disoriented by the distances or simply by the long, flat stretches of dirt, of earth. But as we tour the exterior and interior landscapes of her heartland home, things will be revealed that are imperceptible or invisible to the naked eye—whether from the present or the past. Science turns to art, paleogeology into painting, environmental data into drawings, hospital records and histology into human forms. *Living Downstream* is a layperson's guide to environmental health. It is a short course on how to learn about and track the environmental causes of cancer and to get involved. The subjects, without Steingraber's personal narrative would seem mind-numbing or mundane—the sources and location of ground and surface water, the rise of commercial agricultural and industry, the use of chemicals in postwar central Illinois, the composition of probable carcinogens, the dispersal of toxic substances through the pollution of air and water, the production through combustion of dioxin and furans. And, of course, the initiation of carcinogenesis through endocrine disruption, breaks in DNA, enzyme production, and more. Non-Hodgkin's lymphoma, multiple myeloma, the numbing names of cancer. All these are the findings, the *facts* of science that can set us free, if we could but see, and touch, and feel them. It is why *Silent Spring* begins with a fable. It is why *Living Downstream* opens beneath a starlit sky.

Sandra Steingraber walks through the fields of Illinois with us, calling up each slope, each dip, that is otherwise a seemingly flat, endless expanse of dirt. As it was for Ellen Richards, this is best seen with maps. "Illinois is not flat at all, I would insist, as I unfold geographical survey maps that make visible the surprisingly contoured lay of the land. Parallel arcs of scalloped moraines slant across the state, each edge representing the retreating edge of a glacier as it melted back into Lake Michigan and surrendered the tons of granulated rock and sand it had churned into itself."[11] But even this is not enough to pull us in. "Better than maps is the ground fog on a summer night when I drive you across these moraines and basins. Now you see how the shrouded bottomlands are distinguished from the uplands, the floodplains from the ridges, how the daytime perception of flatness belies a great depth. Out of the car and walking, I encourage you to feel, as we traverse a land that appears to be utterly level, the slight tautness in the thighs that comes with ascending a long grade versus the looseness in our feet that indicates descent."[12]

This sensuousness, the learning from the body, as well as from the mind, is key. So is the imagination, to picture and to feel the facts of abstract science. This metaphorical method prepares us to enter inside, to see beneath the facts, to invite the invisible into our lives. English physician William Harvey discovered circulation inside the body in 1628, Steingraber tells us. Something similar goes on with the water around us and far beneath our feet. Harvey discovered a diffuse net of permeable vessels. "So too in Illinois, a capillary bed of creeks, streams, forks, tributaries lies over the land. . . . this is only the water that is visible. Under your feet lie pools of groundwater held in shallow aquifers—interbedded lenses of sand and gravel—and in the bedrock valleys of ancient rivers that lie below. One of these is the Mahomet, part of a river system that once ran west across Ohio, Indiana, and Illinois." Thousands of tons of debris deposited by melting glaciers buried the Mahomet, which now flows underground. You can stand in a place in Mason County, she tells us, where the Mahomet once met the Illinois. Here, the water is not too far below the surface, so that with heavy downpours, "lakes brim up from under the earth and reclaim whole fields and neighborhoods."[13] The same is true of ancient channels of the Mississippi in Tazewell County, Sandra's girlhood home. "If you could see through the dirt, imagine the dramatic view you would have." We have instead endless fields of corn and soybeans that feed America—especially in processed foods from soft drinks to salad dressing. It is the start, says Steingraber, "of a human food chain. The molecules of water, earth and air that rearrange themselves to form these beans and kernels are the molecules that eventually become the tissues of our own bodies. You have eaten food that was grown here. You are the food that is grown here. You are walking on familiar ground."[14] The ground we eat, we quickly understand, is laced with chemicals introduced after World War II—brought home along with her father. In 1950, less than 10 percent of cornfields were sprayed with pesticides. Now it is 99 percent. Fully 89 percent of Illinois is farmland, and each year some 54 million pounds of synthetic pesticides are applied to the fields on which Steingraber has us walk and imagine the ancient water underneath. It is important to understand all this even if you are an urban dweller.

Only 11 percent of Illinois land is occupied by cities, but it is where most people live. There are approximately fifteen hundred hazardous waste sites there, not counting thousands of "pits, ponds, and lagoons in need of remediation." So, we discover, Illinois injects each year about 250 million gallons of industrial waste into five deep wells that penetrate into bedrock caverns. And, sure enough, "these geological formations are overlain by aquifers and farmland." Our imagining, our feeling of the fields of Illinois and its invisible underground geography and water are aroused so that we can visualize that all those pesticides that are sprayed do not stand still. They evaporate, permeate the soil, dissolve in water, "flow downhill into streams and creeks," enter glacial aquifers and buried river valleys. "They fall in rain. They are detectable in fog." In 1993, 91 percent of Illinois's rivers and streams showed pesticide contamination. From the air and water and earth, they enter the human body and dig deep into our cells. "Some, including one of the most commonly used pesticides, atrazine, are suspected of causing breast and ovarian cancer in humans. Other probable carcinogens, such as DDT and chlordane, were banned for use years ago, but like the islands in preglacial river valleys, their presence endures."[15]

Steingraber, Carson, and the Complicity of Silence

Much of what Steingraber tells us about pesticides and cancer was known in Rachel Carson's time. Even more is known today; we have more tools to trace the toll. The looming question, then, is why do more people not know these facts? Why has relatively little been done about them? *Living Downstream*, in addition to being a gripping account of personal battles with cancer and a lyrical laying out of the consequences of chemicals, is a diary of democracy in action. As Steingraber researches her book, from the Beinecke Library at Yale to the marshes of Maine, she wants to "listen to the voices behind *Silent Spring*"; she ends up thinking about silence. America is a democracy, she says, with guarantees of free speech "carved into the heart of our legal system." No one is carried off by secret police at night, nor are passages in books "blacked out by a censor's invisible hand." Instead, we have subtle codes of silence, unspoken agreements in the workplace, or "a family secret that everyone knows but does not discuss."[16]

Rachel Carson was interested in these silences and wanted to interrupt them with her writing. The dangers of DDT and other pesticides had been known and discussed from when they were first synthesized. Her friends and government colleagues were debating them fiercely as early as 1945 when Carson first considered writing about the problem of pesticides. But the data and information was not "spirited away in the middle of the night." Instead, it "remained soundproofed in internal documents and technical journals . . . research was sorely underfunded . . . government officials turned deaf ears to bearers of bad news."[17] Silence in a democracy is the true subject of *Silent Spring*. It can be read, Sandra says, as an account of how "one kind of silence breeds another." Rachel Carson warns, for example, that the failure in Iroquois County, Illinois, in a useless, protracted chemical war against Japanese beetles accomplished nothing. But it did leave residues of dieldrin, undiscovered, undiscussed, throughout the water and the soil. Such silence cascades through ecosystems. Robins eat poisoned grubs and worms and die themselves. In Iroquois County, the result, among others, is dead ground squirrels, their mouths full of dirt, having "gnashed at the ground as they died."[18]

Perhaps most important, Steingraber wants us to see, is the silence of scientists themselves, their failure to speak out publically about what they have found. Rachel Carson did not respect such cowardice; she wrote to Dorothy Freeman, "I told you once that if I kept silent I could never again listen to a veery's song without overwhelming self-reproach." Then she quoted Abraham Lincoln, "To sin by silence when they should protest makes cowards out of men."[19] It is why we must recall, says Steingraber, that Rachel Carson was political. She spoke to Congress, to the public, raised her voice in protest. It is why when her voice fell silent with her death from cancer, others must carry on. Sandra's father and Rachel Carson are not just interesting medical cases. We are meant to feel their loss. The same is true of Jeannie Marshall, Steingraber's close friend from Boston. Marshall joins her on field trips and research; they literally trace the path of Rachel Carson. But Sandra is also at Jeannie's side at each step of her diagnosis and the spread of cancer. The procedures, the agonies, the arrogance and distance of some doctors, the search for meaning, are all carefully described. There is no silence, cancer is not a secret, as it was kept and so little discussed between Dorothy Freeman and Rachel Carson.

Feeling the terrible loss of a friend like Rachel Carson or Jeannie Marshall is essential if the silence of science and of statistics is to be broken. And Marshall's death serves as poignant, pointed introduction to the meaning of modern cancer mapping. We no longer imagine mere dots upon a map or chart. We are meant to see and feel the lives of those who are counted, as did Rev. Whitehead in East London, Alice Hamilton in Illinois, and Ellen Richards who broke the silence of an American Public Health Association meeting in Boston by recounting the deaths of schoolchildren from the toxic hazards found in the city's schools. She called their deaths the fault of the city elders who failed to act, calling it murder as surely as with a gun.

Seen in such searing, personal terms Steingraber's lucid explanations of science and her own search for causes are unforgettable. The tangled gobbledygook of environmental policy and science is cleared away. Thanks to laws like TOSCA (the Toxic Substances Control Act) and FIFRA (the Federal Insecticide, Fungicide, and Rodenticide Act) and other acronymic abominations, we have tools such as data from the Toxics Release Inventory (TRI). Large American companies are required by law to measure and report the kinds and amounts of dangerous chemicals emanating from their factories. Not all companies are covered, not all data is complete, but from it you and I can find out what is being put into our own townships and terrain. Steingraber does this for her own hometown of Pekin, not far from Peoria, the main street of archetypal America. Once the chemicals are publically reported, we can find which ones are known or probable carcinogens. Then from cancer registries, we can find out what cancer cases and deaths have been recorded in places like Peoria or Pekin. Not all cancer cases are recorded. Registries are not uniform or kept in every state. But there is often enough information to spot a problem or a link. The results then form more modern maps, with colors and computers, than those so painstakingly penned by Ellen Richards or John Snow. Making the data personal, visual, emotional, as well as scientific, we learn from Steingraber how cancer rates (incidence and prevalence) are calculated, how we can see the changes in an ethnic population when they immigrate to a new country and develop its diseases. This happens to Japanese women who show higher rates of breast cancer when they move to the environment of the United States. We see how long-lived molecules make it to the far north through wind and distillation, entering the food chain and changing the lives of far-off women. We also see why animal testing is used to verify the effects of chemicals when experiments on humans are ethically ruled out.

Sandra Steingraber, Cancer Survivor

Finally, we are with Steingraber when she reveals her own feelings and fears as she is diagnosed in college with the bladder cancer that changed her life, giving her the calling to follow Rachel Carson. Even in the 1970s, her roommate moves out, fearful of catching the disease. Steingraber recalls opening her dormitory room door and seeing the bare mattress. "I became secretive and territorial. I staked out a favorite stall in the women's bathroom. In my return every third month to the hospital for cytoscopic checkups, cytologies, and other forms of medical surveillance, I told no one where I was going." Her checkups become annual, but never leave her. Like breast cancer, "bladder cancer can recur at any time, lying quiescent for years— sometimes decades and then reappearing inexplicably." What she means to say by recounting her own hospital checkups with their cancer lingo is: "First, even if cancer never comes back, one's life is utterly changed. Second, in all the years I have been under medical scrutiny, no one has ever asked me about the environmental conditions where I grew up, even though bladder cancer in young women is highly unusual."[20]

Living Downstream eloquently allows us to learn about and to feel a powerful case for the environmental causes of cancer. We learn that most cancers are not genetic, nor simply a case of lifestyle choices. But the emphasis in the cancer establishment, as in Rachel Carson's day, remains focused on cures and not prevention. Despite *Silent Spring*, the incidence of a number of cancers is on the rise, and too much silence remains on the subject. It is why *Living Downstream* concludes with detailed information about environmental and health organizations and sources on how citizens can exercise their right to know and demand change. It is more than a call to arms. It is a call to organized, enlightened activism.

In her next two books, Sandra Steingraber follows most decidedly in the direction of Ellen Swallow Richards. Ellen Richards shocked the public health and medical establishment by saying that those failing to clean up the health hazards in Boston's public schools were no different from murderers. A backlash began, and she and other pioneers of public health saw progress grind slowly to a halt. Increasingly, Richards turned to lecturing, writing, and education, focusing on women and their special sphere as guardians of the home. She founded the field of home economics, but not the pabulum it became after World War II and the baby boom. She believed that if ordinary women, housewives, mothers, and workers could understand the need for clean food, air, water, and the economic and corporate forces that prevented them, they could, in effect, change the world. Sandra Steingraber's work after *Living Downstream* is somewhat similar. It is as if her father's love of organic gardening and her own observations of the environmental causes of cancer have come together. Steingraber turns her scientific eye and poetic pen inside herself while she is pregnant with her first child, Faith. Then, after the birth of her second child, Elijah, she follows his growth and the environmental hazards and health threats that he faces in playgrounds, schools, and neighborhoods. A healthy life is a fundamental human right for ecologists like Carson, Williams, and Steingraber. But where Terry Tempest Williams leads us finally to connect to issues like genocide, Sandra Steingraber focuses in on American children and our communities. She has us wonder why good parents worry about alcohol or abuse, report cards and reading levels, but not about the potent, present dangers in the contemporary environment.

The result is two books, *Having Faith: An Ecologist's Journey to Motherhood* (2001) and *Raising Elijah: Protecting Our Children in an Age of Environmental Crisis* (2011).[21] Both draw us into unseen worlds to make real the marvels and the mysteries of environmental science. But despite the domestic topics, there is no separate sphere here for women or for children. For the modern woman, as for modern ecologists, the waters of the womb, the particles on the playground know no boundaries. The hazards they present are socially constructed. They are everyone's burden to bear. Steingraber's

FIGURE 5.1   Sandra Steingraber with Faith and
Elijah. © 2013 by Carrie Branovan for Organic
Valley.

strength is in her openness and honesty about her own fears about moth-
ering, breast feeding, bringing up boys, and the mistakes that she and Jeff,
her artist husband, often make. There is no guilt tripping or scolding here.
Sandra Steingraber is like having a roommate or relative you can confide
in, but who knows science and sex and delivers with common sense and a
sense of humor. She is particularly good at reminding parents that the entire
burden of protecting their kids should not be theirs alone. Being a careful consumer, a safe shopper, is important. But joining with others to prevent harmful environmental exposures—local, national, and global—will
require active engagement as a citizen as well. Unless we change our society
and our social norms, our pollution can reach the polar regions; furans from
a factory far away can harm a woman's fetus.

The Placental Barrier Is Broken—The Personal Is Political

This interaction between environmental harms and the personal and the political is key for Steingraber. Most mothers are unlikely to consider whether the EPA has anything to do with the embryo inside them. But it does. As Steingraber puts it, "When I became pregnant for the first time, I realized, with amazement, that I myself had become a habitat. My womb was an island ocean with a population of one."[22] This realization leads directly to another, that "protecting the ecosystem inside my body required protecting the one outside." This idea was held by the ancients, like Aristotle and Hippocrates, who mistakenly thought the placenta was the place where a mother's blood went directly into the fetal umbilical cord. They were wrong, but had the right idea. Substances pass from a mother through the placenta to her unborn child. Even in Carthage, newlyweds were not allowed to drink on their wedding-night to prevent any damage if conceiving. But later medical experiments seemed to indicate that a mother's blood does not mix with that of the fetus. This gave rise to the notion of the placental barrier—a kind of impervious wall between mother and her unborn child. The impermeable placenta held on in medical schools and teaching well into contemporary times, even though by the 1950s there was plenty of evidence that animals had offspring with birth defects as a result of environmental exposures, that exposure to X-rays, pharmaceuticals, and even chemicals somehow might be harmful to fetuses in the womb.

It took four major tragedies to dispel the myth of the placental barrier and, hence, of the danger of environmental toxins. The first was an outbreak of German measles, or rubella, a mild form of measles, which caught the attention of an Australian ophthalmologist, Dr. N. McAlister Gregg. In a 1941 paper, he linked blind babies with congenital cataracts and other diseases, such as heart defects, with their mothers' exposure to rubella during pregnancy. Yet it took a major global rubella outbreak in 1964 that maimed more than twenty thousand infants born in the United States to lead to the first vaccine in 1969. Gregg's paper was later hailed for its insights, Steingraber tells us, but not for its equally important warnings about other "toxic influences . . . known to be transmissible transplacentally."[23] That would include the second major tragedy from environmental toxins from my youth—thalidomide. Developed in Germany in 1953 to be an anticonvulsant (it proved ineffective), thalidomide was then heavily marketed as a safe sedative and, finally, as an aid for morning sickness among pregnant women. It was not safe. More than eight thousand children in Europe and Canada were born with severe birth defects, especially "reduction limb deficit," or dwarfed and missing limbs, including "phocomelia," a condition in which limbs looked like tiny flippers. There were also untold miscarriages and stillbirths. The tragedy is almost unimaginable. Steingraber puts it more personally. "The damage was more than physical. The birth of these babies wrecked marriages, impoverished families, and crushed mothers under the weight of relentless guilt."[24] Thalidomide was marketed worldwide, despite clear danger signs in forty-eight nations before global tragedy unfolded. But not in the United States.

In 1960, a pharmaceutical company filed for permission to make and market the drug here in the United States. Only the cautious investigations of Dr. Frances Kelsey, new at the time to the Federal Drug Administration (FDA), held back approval. She recalled her own work with malaria and how embryos could not metabolize quinine, the antimalarial drug, in the same way as adults. And she recalled the story of rubella. She kept asking questions and demanding more answers until the stories of limb malformation came out in Germany and England. Dr. Kelsey received a distinguished service award from President Kennedy. But there are two more things Sandra Steingraber wants us to remember from this story of a heroic female doctor. One is that we cannot wait until we know everything about a potentially harmful substance before we hold back its use. The other is that the timing of fetal exposure is as important as the dose. This observation runs counter to the long-held axiom in schools of medicine and public health that "the dose makes the poison." This means that there are presumably safe levels for some chemicals, or, that there is a threshold that must be crossed before a toxin is considered dangerous. Tiny exposures are unlikely to cause harm. And the larger the dose, the worse the outcome. Thalidomide showed that neither was the case. It was the timing of exposure that mattered. If the pills were taken between days 35 and 37 of pregnancy, the baby would be born without any ears. If between days 37 and 39, no arms. And so on.

Minamata and Mercury Poisoning

But the lessons of thalidomide are now far behind us, as are the results of another modern toxic tragedy—Minamata. Steingraber asks her husband, Jeff, if he remembers thalidomide. He does. As they did with me, the deformations, the flippers made a deep impression. But now two-thirds of those under forty-five years of age do not recognize the word. Sandra asks if he recalls Minamata. Jeff says no. How about the famous photograph "of a Japanese mother bathing her paralyzed daughter?" Jeff, an artist, recalls the 1975 *Life* magazine photo by W. Eugene Smith. "It was black and white and darkly lit. It was composed like Michelangelo's *Pietà*, but it was also a baptism. That's what I remember."[25] As it turns out, Minamata is the home of Chisso, a Japanese manufacturer of acetaldehyde and vinyl chloride, both components of plastics. The plant used mercury as a catalyst for its process. The run-off containing mercury went into Minamata Bay and from there was transformed by bacteria into the organic compound methyl mercury. Methyl mercury, a potent neurotoxin, then entered the food chain and ultimately poisoned the local population. The company ignored evidence, stalled, denied harm for many years, claiming the "disease" had other causes. But following her theme of the dangers of silence, Steingraber tells us that ultimately a Chisso company doctor, Dr. Hajime Hosokawa, became aware that Minamata disease was related to the discharges from his factory. It was not an infectious disease after all. He had proved this by feeding experimental cats with Chisso sludge. They developed Minamata disease. But unlike Drs. Gregg or Kelsey, Dr. Hosokawa kept silent, as did his bosses at Chisso. Chisso kept using mercury, and the disease went on until lawsuits were filed by citizens, demonstrations were carried out, and protestors were beaten — including the *Life* photographer Eugene Smith. As Steingraber says, "In the end, it was citizen activism and photography, and not the slow accumulation of scientific knowledge, that awakened awareness about the ecology of methyl mercury."[26]

The Damage of DES

The final tragedy that gave rise to the modern concern for exposures to toxins during pregnancy is that of the use of DES by pregnant U.S. mothers, including the one who gave birth to filmmaker Judith Helfand. Her documentary, *A Healthy Baby Girl*, has been widely shown on campuses and both Sandra and I have spoken alongside Judith. She, too, is remarkably candid about her own life as a victim of DES, diethylstilbestrol, a hormone once given to pregnant women to prevent miscarriages. We now know that it causes cancer and infertility in their daughters. In Helfand's case, it is cervico-vaginal clear cell adenocarcinoma. As she describes it in her film, it means she has lost her uterus, cervix, and the top third of her vagina. Again, warning signs were everywhere, but ignored. Synthesized in the 1930s, DES fattened livestock and poultry and mimicked estrogen. It was used to suppress breast milk for bottle-feeding mothers, and, finally, to prevent miscarriage. But animal studies in the 1930s already indicated that DES caused breast cancer in mice as well as malformations in reproductive organs. Later studies with humans in the 1950s showed that it actually increased miscarriages. But it was still prescribed for another dozen years. In the late 1960s, the medical staff at Massachusetts General Hospital, baffled by a cluster of clear cell cancer of the vagina, repeated the rubella action. They listened to the mothers of the victims; they recalled taking DES during pregnancy. By 1971, Dr. Arthur Herbst and his colleagues presented a paper with the truth about the dangers of DES. As Steingraber says again, "timing of exposure proved as important as dose." This introduction to the main historic events leading to the modern understanding of the dangers of environmental toxins sets the stage for her own experience of pregnancy—the lessons we all can draw from the failure to anticipate and prevent harm, from remaining silent, and from ignoring clear warnings of danger in the emerging scientific and medical literature. No one has suppressed the information we need to change things and protect our children. But it is necessary that we understand and know the basics about environmental health and then take action.

Amniocentesis Tests the Environment

In her chapter called "Egg Moon," for example, Steingraber relives with us all the doubts, fears, advice, and procedures connected with amniocentesis. She is on the borderline, at age thirty four, whether or not to have one. We learn not only the biology, the reasons and the risks of amniocentesis; we learn that the energy around the procedure has been focused almost entirely on finding *genetic* abnormalities, even though the majority of birth defects are *not* attributable to genes. Steingraber hopes we will someday wonder with her why we "ritualize amniocentesis as a rite of passage for pregnant women, as though chunks of DNA were the prime movers of life itself. As though pregnancy took place in a sealed chamber, apart from water cycles and food chains."[27] As she wrote, only one study of environmental contaminates in the womb had ever been done, and it found organochlorine pesticides, including DDT, in one-third of the samples. Clearly, we have the science, the tools to learn more. What we face is the silence that Rachel Carson warned about.

This problem is exacerbated by the lack of sufficient data on birth defects themselves, whether or not they might be caused by environmental toxins. In the United States, only California and Texas have decent registries for birth defects, but even there, the causes and links to the environment are hard to track. Steingraber finds this an amazing discovery, because "birth defects are the number one killer of infants in the United States, and, by any count, the prevalence of birth defects remains high despite other improvements in the health of infants and pregnant women."[28] Among American infants, 120,000 are born with major deformities each year. Twenty-one babies die as a result of these defects each day.

I became aware of the horrors of severe birth defects and the difficulties of tracing them to the environment in south Texas, along the border where Brownsville and the Mexican city of Matamoros meet. I met with pediatrician Dr. Carmen Roca, who had called attention to the problem, along with members of the community on both sides of the border. In a short period, a number of babies had been born without parts of their brains, or in some cases, no brain at all. The condition, called anencephaly, can be caused by exposure to toxic substances or by a lack of folic acid in the prenatal diet—or both. Mothers in the area were convinced that the local *maquiladoras*, small, unregulated manufacturing facilities in Matamoros, were the cause. The Texas Natural Resources Center and the Agency for Toxic Substances Disease Registry (ATSDR) combined forces to do a study. Almost predictably, given the difficulty of tracking exposures, the experts announced that lack of folic acid must be the cause. Toxic chemicals would need further study. The mothers were, of course, outraged, and they became only further deeply skeptical of whether scientists and academics could show any sympathy or cultural understanding. What is clear, however, as Steingraber reminds us, is that heredity and lifestyle are not the problem. And from other studies we do know that the chances of anencephaly are greater among the children of fathers who handle toxic chemicals at their jobs, such as painters or pesticide applicators. There are similar problems in tracking the growing rates of hypospadias, a condition where the urethral tube of a baby boy does not emerge at the tip of the penis allowing normal urination. It is malformed to appear somewhere else along the shaft of the penis or, in severe cases, at the scrotum. These and other problems have been linked to toxic exposures, but the data and registries are incomplete.

Why Not Track the Chemicals?

Not one to give up, Steingraber attacks the problem of birth defects the other way around. If birth registries are of little use and do not count malformations in aborted fetuses or stillborn babies, how about data on the release of potential teratogens (chemicals known to possibly cause birth defects)? The answers are still impossibly hard to find. Of some eighty-five thousand synthetic chemicals produced in the United States, some three thousand are produced in high volumes. More than three-quarters of these have had no toxicity testing at all. Of seven hundred found in widely used consumer products, nearly half lack any information on possible developmental toxicity. Some of these chemicals are covered by the Toxics Release Inventory, which Steingraber relied on for her hometown hunt for cancer in her first book, *Living Downstream*. But TRI data covers only about 5 percent of all chemical releases in the United States. Even still, the list of fetal toxicants for the year in which *Having Faith* was written stands at 989,700,000 pounds. They are given off by chemical manufacturing and in the production of paper, metal, rubber, and electric power. For Illinois, which ranks fourth in toxic emissions, fetal toxicants add up to 39,500,000 pounds.[29] Other countries, however, especially in Europe, do track birth defects and teratogens more carefully—including abortions and stillborn babies. The results indicate that chemical exposures for people living near landfills and hazardous waste sites are associated with "adverse reproductive outcomes." Common in these places are many organic solvents, which contain a group of common chemicals associated with birth defects, including kerosene, acetone, benzene, xylene, and toluene. Others, like dry-cleaning fluids, carbon tetrachloride, and TCE (trichloroethylene), also pose a threat of birth defects and other problems.[30]

The workplace, or "the dangerous trades," as Dr. Alice Hamilton first revealed, is also a source of serious exposure to fetal toxicants. And the problem is not just with old-fashioned industry and men exposed to huge vats of molten metals. Contemporary jobs where women work are full of such risks, whether in health care, office work, housecleaning, dry cleaning, printing, graphic design, and more. Steingraber reports that studies show that, in Sweden, women working in laboratories have babies with more defects; women doctors in England and Wales who work with anesthesia while pregnant give birth to babies with more heart and circulation problems and have more stillbirths.[31]

Steingraber on Men's Reproductive Health

Because Sandra Steingraber consistently seeks to involve the largest number of readers and to offer a human face and sympathy amid potential tragedy, she reflects on her own thoughts about fetal health. She talks with her husband and reminds concerned readers that birth defects caused by exposures to chemicals are not restricted to women. Most birth defects are not related to male exposures in the workplace, but there are important exceptions and increasing evidence that men are not free from "the birth defect lottery." Among men, painters have more children born with anencephaly and heart defects. Farmers have more kids with cleft palates and lips, as do firefighters. And the list goes on. But exposures can be found at or near home as well. Sandra muses about dioxin and Agent Orange in Vietnam when she and Jeff stroll by a lawn with little flags alerting people that it has recently been sprayed with pesticides.[32] As the couple prepares to move back to the Boston area from central Illinois, where Sandra has been a writer in residence, we are reminded that in Illinois, the groundwater and drinking water show traces of atrazine. But the answer is not to avoid playing on lawns or to drink bottled water (which is not regulated at all), but to think about the overall environment that a fetus or a female firefighter or a father is exposed to.[33] As Sandra and Jeff discuss how disturbing the possibilities of birth defects can be, why people may want to avoid the subject, and how free of discussion most baby books are, they consider public art projects and other ideas to break through the silence on the subject. Baby books warn, and mothers and the public are now aware of the dangers of coffee, alcohol, and cat feces during pregnancy. But these all urge private responses, not public ones. "It's pregnant women who have to live with the consequences of public decisions. We're the ones who will be raising the damaged children. If we don't talk about these things because it's too upsetting, how will it ever change?"[34]

The horrors of severe birth defects first brought the attention of the world to the connection between chemical exposures from thalidomide and DES and their harm to human fetuses. But having shown that the timing of exposures, not necessarily the size of the dose is critical, Stein-graber explores the subtler effects on brain development, intelligence, and behavior that are far harder to measure and notice; these can come from even tiny exposures, including ones that even the most conscientious and alert mother may not be able to control. Here, as she discusses lead, mer-cury, and the newer persistent organic pollutants (POPs) that so concerned Rachel Carson in *Silent Spring*, Steingraber directly descends from Carson. But she is also closely related to Ellen Richards and Alice Hamilton.

Although Richards ultimately turned to home economics, writing, and public lecturing after being driven from the academy, she never urged that women see their own plight as isolated or merely take private, personal steps to protect themselves. Her goal for home economics was to educate, alert, and involve women in science and the interaction between the household and community environment with the larger world of pollution and public policy. So while swiftly summarizing the role of American industry in deny-ing the harmful nature of their products and waging massive PR and lobbying campaigns to keep them from being regulated and having toxins removed, Steingraber reminds us that progress has been made, but there have been no final victories. The story of lead, for example, "seems like a story of science triumphant over ignorance. Lead paint was banned in 1977 and leaded gas phased out and finally banned in 1990."[35] The result is that levels of lead in the blood of American children plunged by 75 percent between 1976 and 1991. But that is not the end of the story. Lead is still omnipresent in the United States, and lead poisoning remains a problem, even if much reduced. Lead is found in cosmetics, lipsticks, and dyes—to such a degree that one in twenty contemporary American children suffers from lead poisoning. In Somer-ville, Massachusetts, where Sandra and Jeff live at the time she is writing her book, residents are advised not to grow root vegetables in their gardens because of lead in the soil.[36] Both Ellen Richards and Alice Hamilton warned and campaigned against lead. But, at each turn, companies like General Motors and the Ethyl Corporation and the National Lead Company (which dreamed up the Dutch Boy brand and symbol for paint) won exceptions and delays, and denied the science. As a boy, one of my treasured possessions was the little Dutch Boy hat I got with my dad at the paint store. While literally brainwashing a generation (lead is a neurotoxin that lowers IQ and impairs mental performance), the National Lead Company fought product label-ing, not to mention bans; brought lawsuits; and finally, when the danger was undeniable, blamed the children (and their families who somehow didn't prevent it) who consumed lead paint chips that flaked off onto floors and rugs and throughout older housing and apartments. One industry representative went so far as to suggest that the problem was not that lead paint makes children stupid, but that stupid children eat lead paint.[37]

But for Sandra Steingraber, the personal is always the political, and environmental health history, scientific studies, and statistics are never sufficient. It is far more memorable, and the meaning becomes eminently clear, when we learn that her husband has highly elevated lead levels from his work as an artist; Sandra and Jeff's apartment also contains lead paint. They agonize about what to do. Neither makes much money. Moving will be difficult, and the young parents-to-be need two incomes. They decide to stay in the apartment (though regret it later when new studies reveal that even smaller and smaller exposures to lead are still dangerous). Jeff very regretfully decides that it is best that he give up his art business. As he puts it, "Don't grow our own root vegetables. Quit a job I like. How come we're always the ones that have to do the abstaining?"[38]

Precisely. The most compelling points of *Having Faith* are that we modern Americans (and our counterparts around the world), no matter how educated, careful, and conscientious, are exposed to and carry body burdens of toxins and pass on tiny amounts to fetuses through breast milk and across the placental barrier. We are all connected to the external environment. Personal decisions are important. But understanding and acting on the outside world is even more so.

Steingraber takes us through her entire pregnancy and gently explains, demystifies, and warns. *Having Faith* is a wonderful baby book, modern pregnancy guide, and introduction to the field of children's environmental health through the experiences of a poetic, sympathetic biologist. But just as toxic chemicals can cross the placental barrier, Steingraber breaks down long-established boundaries—the personal and the political, the womb and the world, science and sensitivity, expertise and activism.

Sandra Steingraber in Action

Toward the end of *Having Faith*, after Faith is born, and after a brilliant exposition of why breast feeding is important and healthy, despite the presence of some toxins, Steingraber goes to Geneva, Switzerland. She is there in September 1999 for the Third Session of the Intergovernmental Negotiating Committee for an International Legally Binding Instrument for Implementing International Action on Certain Persistent Organic Pollutants. It is better known as the POPs Treaty negotiations; I was there, too, as the head of Physicians for Social Responsibility, with key staff, Susan West and Karen Perry, who led the international NGO efforts to get this close to a treaty. It is because of the treaty that emerges from Geneva that today's right-wingers and polluters falsely attack Rachel Carson as responsible for genocide in Africa. Carson would have been a revered figure had she lived to be in Geneva at ninety-two and not died of breast cancer linked, among other things to POPs, including DDT. And like PSR, Rachel Carson would surely have approved of the pragmatic compromise that emerged in the treaty now known as the Stockholm Convention. Thanks to the work of PSR and the International POPs Elimination Network (IPEN), DDT was not immediately banned, as a few cranks have contended. It was given a public health emergency exemption for use in the event of serious malaria outbreaks; it was to be phased out as alternatives emerge. Those who were there who tried to stir up enmity toward American environmentalists and public health experts were really out to scuttle a treaty, not improve it. And there is no real evidence that saving Africans was on their minds or that allowing DDT to be used forever would help. It would, in fact, lead to a variety of long-term health problems, including breast cancer. And so, serious science and Rachel Carson's spirit were very much present in Geneva at the POPs negotiations.

In Geneva, Steingraber wants to make a difference with her invited presentation to United Nations delegates from 122 nations. She frets and worries while her husband and daughter go to art museums, outdoor cafés, and the Swiss National Circus. She attends IPEN meetings and goes to workshops and the women's caucus group. She is scheduled to present the science of the reproductive health effects of POPs along with two esteemed male colleagues. They will use graphs, charts, references. But she has also been told that she can speak more specifically and personally, if she wishes, on breast milk contamination. Even as she heads to the forum to speak, she is still unsure how best to "strike the right balance between the intimate and the empirical."[39] She wants to speak as a nursing mother and as an ecologist. As Steingraber arrives, she needs to pump breast milk. But rather than throw it away, it suddenly occurs to her that most of the delegates have never seen the stuff. "Breast milk may be the most contaminated human food on the planet, but it is still Holy Water to a mother." As she begins to speak, Sandra sends a jar of her breast milk around the room to the startled delegates. "Some study it closely. Some avert their eyes. Some smile with recognition." And then she talks about the food chain and how it is that mother's milk around the world has come to be contaminated by POPs.[40] Her presentation and her science are impeccable; the impact of breast milk for an infant undeniable.

She ends by describing the activism and the advocacy of NGOs, which eventually will bring about a POPs treaty, signed in 2001 and called the Stockholm Convention. It includes the precautionary principle, immediately bans eight pesticides worldwide, and restricts two others. It calls for an end to PCBs in electrical transformers by 2025. Dioxins and furans are to be reduced right away and eliminated where feasible. DDT, the chemical that researchers and Rachel Carson worried about as early as 1945, and that is still in use almost forty years after *Silent Spring*, is to be employed only in emergencies and replaced over time by substitutes for which funds are provided to poor nations.[41] It is a lovely testament to the work of women writers, scientists, activists, and advocates. But it is still not enough. The Stockholm Convention, signed by the United States in Stockholm in 2001, has yet to be ratified by the U.S. Senate. In *Having Faith*, Sandra Steingraber calls for an ongoing campaign by women for the precautionary principle and an end to the production and use of toxic chemicals and ratification of the Stockholm Convention. She is still at it today.

Steingraber and *Raising Elijah*

Sandra Steingraber brings environmental health up to the present in her book *Raising Elijah: Protecting Our Children in an Age of Environmental Crisis*. In the ten years after *Having Faith*, the environmental crisis has deepened, especially as global climate change accelerates and has yet to be seriously addressed.

Here Steingraber explicitly links the two branches of environmental threats—from the burning of fossil fuels and from the spread of toxic chemicals—to the planet and to organisms, including humans. As in

Silent Spring, her concern is not just with nature—the robins that will fall silent—but with humans who will also suffer. They are inextricably linked. "Biologist Rachel Carson first called our attention to these manifold dangers a half century ago in her 1962 book, *Silent Spring*. In it, she posited that 'future generations are unlikely to condone our lack of prudent concern for the integrity of the natural world that supports all life.' "[42] As she puts it, "Since then, the scientific evidence for its disintegration has become irrefutable, and members of the future generations to which she was referring are now occupying our homes. They are our kids."[43] The preciousness of life, the stuff of Albert Schweitzer and the spirit of Rachel Carson, is central to *Raising Elijah*. And so is breaking the silence that permits injustices to roll on. A reverence for life means living boldly and bravely, speaking out, having courage. Her son, Elijah, is not named after the prophet of old, at least not directly. It is Elijah Lovejoy, the martyred abolitionist editor, killed by a mob in Alton, Illinois, who has inspired Steingraber. She begins with the Reverend Lovejoy, whose presses have already been destroyed, riddled with bullets in 1837 by "a mob intent on silencing the man once and for all." They succeeded, but not for long. The horrible news spreads far and wide. Antislavery societies swell with members. Edward Beecher, president of Illinois College and Lovejoy's friend, becomes an ardent abolitionist and inspires his sister, Harriet Beecher Stowe. Elijah's brother, Owen, opens his home to runaway slaves on what becomes known as the Underground Railroad. He goes to Congress and makes friends with an up-and-coming Illinois politician named Abraham Lincoln. Steingraber says simply, "These facts impressed me as a child."[44] She is impressed not only with the bravery of Elijah Lovejoy, but that he wrestles with his conscience; he worries if he is doing the right thing. At thirty-four, Lovejoy has a one-year-old child and a pregnant wife in the bed beside him. Publicly and personally calm in the face of death, he writes to his mother, "Still I cannot but feel that it is harder to 'fight valiantly for the truth' when I risk not only my own comfort, ease and reputation, and even life, but also that of another beloved one. . . . I have a family who are dependent on me. . . . And this is it that adds the bitterest ingredient to the cup of sorrow I am called to drink."[45] Lovejoy was calm in the face of slavers and thugs. But his language was fierce for those who silently stood by. Steingraber hopes that her son will admire the Elijah Lovejoys of this world and learn that his mother wrestles with her own fears and family life. But finally, she calls for "outspoken, full-throated heroism in the face of the great moral crisis of our day: the environmental crisis."[46] It most affects children, and so she addresses her book first to families.

Raising Elijah is written and follows events over the first decade of Elijah's life. Through its stories and statistics, it draws on the work inspired by a growing twenty-first-century environmental health movement and the science that has evolved along with it. Her first book focused on the environmental causes of cancer; her second on the effects of small doses of toxic substances on fetuses and child development—cancer, birth defects, learning and behavioral problems, and more. Her third book shows what we have learned about the harm from minute doses of chemicals—even in family food or familiar consumer products. And, of course, what we know from the science supporting the dangers of global climate change that became conclusive with the Fourth IPCC Climate Assessment of 2001—issued the year that baby Elijah Steingraber was born. The ordinary things of American life, automobiles, electric lights, farms, and factories now, too, must change. In this call for change and a focus on families and consumers, Steingraber picks up another theme from Ellen Swallow Richards—a growing consumer protection movement. In 1879, Swallow was asked by her mentor, Maria Mitchell, to address some three hundred women in Poughkeepsie at one of the growing number of women's clubs. She did, indeed. Swallow used the occasion to call for educated consumers to apply real knowledge, real science, and education to protect their families and society from a variety of ills. She condemned unfair pricing and products unfit for human consumption; she called for citizens to learn about and organize against unsanitary living conditions, polluted air, contaminated water, poor nutrition. "Can a woman know too much about the composition and nutritive value of the meats and vegetables she uses . . . the effects of fresh air on the human system or the danger of sewer gas or foul water? We must show [that] science has a very close relation to everyday life . . . train [women] to judge for themselves."[47]

Childhood and the Environment

And so Steingraber takes us through familiar scenes from childhood as Elijah and Faith grow up and she and Jeff try to educate them, keeping them both healthy and secure. Good parents that they are, Jeff and Sandra search extensively for the best nursery school for Faith in their new home in Ithaca,

New York. Soon Faith is excitedly playing with other three- and four-year-olds on a marvelous playground that features an extensive, imaginative set of climbing decks and stairs like ships of old. It is perfect for active, creative children. But the happy story soon turns sour. An inquisitive, alert chemist in Connecticut, David Stilwell, has become aware of studies that link obscure neurological diseases among workers in wood treatment plants: a government worker assembling a picnic table in an unventilated area, a Wisconsin family that falls ill after burning pressure-treated lumber in their home wood stove. All have been exposed to arsenic. Stilwell begins to crawl around and measure underneath backyard decks made with pressure-treated wood. Result? Elevated levels of arsenic in the soil that exceed cleanup standards for toxic waste repositories. He also finds that he can wipe arsenic from his hands simply after running them along the poles of equipment found in playgrounds.

Once again, the history of environmental and occupational health springs to life in Steingraber's pages. As with many hazards, the use of arsenic-treated wood began as a miracle reform. Sonti Kamesam, an engineer in India, discovered that wood treated with a combination of copper and arsenic prevented rot. The new process soon saved the lives of numerous coal miners from collapsing shafts as a result of rotten beams. Treated beams were then tested by pounding them into fields infested with termites. They all stood the test. In 1950, Bell Telephone applied for permission to use chromated copper arsenate (CCA) on its widespread poles. Arsenic was already a well-known poison in acute exposures. But by 1950 it was suspected of causing skin tumors, though its ability to cause bladder and lung cancer was not yet known. CCA remained a specialty product, and decks and picnic tables and the rest were made of redwood, which is naturally rot and insect resistant. As the price of redwood soared in the 1970s, industry turned to cheaper plantation-grown stands of pine. But the cheaper stuff rotted out and was eaten by bugs. And so wood pressure-treated with CCA was born and quickly caught on. By 1978, the EPA became concerned about the cancer risks of CCA and began a review with an eye to banning CCA from consumer use. But the review took about ten years. By then, the manufacture of treated wood had grown by some 400 percent. Now it was used in picnic tables, gazebos, even children's playgrounds. In 1988, the EPA reached its decision. CCA was reregistered as a pesticide for wood; its use could continue, but it was recommended that warning labels be attached to consumer goods. Industry objected, and so it was agreed that having retail stores that carried treated wood place tear-off pads of information sheets nearby would certainly suffice.

Then in 2001, a newspaper reporter named Julie Hauserman, a latter-day Elijah Lovejoy, wrote an investigative story called "The Poison in Your Backyard." She tested for arsenic in playgrounds and in landfills loaded with treated wood. The poison was well above safe levels and came off easily on the hands. An average five-year-old playing on CCA-laced equipment could exceed "acceptable" levels for cancer risk within two weeks. The same is true for household decks. And Steingraber discovers that both her daughter's playground and her new home in Ithaca feature arsenic-treated wood. So she and a biology teacher friend test the playground, reveal the dangers, and call for the removal of the poison-laden gangways, decks, and ladders. They lose. The other parents decide that paying for removal and new kinds of playground equipment is simply too inconvenient and too expensive. Sandra and Jeff and three other families reluctantly take their children out of the school it has taken them so much trouble to find. We further learn that industry has fought every attempt to restrict or remove or ban its products up until this day. We are left with things like inadequate tips in books such as *Pediatrics for Parents* that suggest you should wear gloves and a mask when sawing or sanding such wood, and that you seal the wood annually, launder separately any clothes that were worn while working with the wood, and not allow children or pets under a treated-wood deck. Sandra is furious at it all, as a mother with small children should be. There is no doubt that some number of children will die from cancers acquired from lying on arsenic-treated playgrounds and backyard decks, perhaps as many as one in ten thousand, a rate that exceeds that of childhood drowning. The difference is that drowning is sudden, tangible, and tragic. Arsenic needs twenty to forty-five years to cause its cancers. "By then nobody remembers daily recess in the nursery school castle or the paper doll parties on the deck. . . . We can't name the dead, and we can't name those responsible for the dead. But the suffering will be real." It will take more articles and books and speeches and a movement to put an end to such insanity. In the meantime, she removes her backyard deck and will have Faith and Elijah play in living trees. "They would not necessitate disposal in a hazardous landfill. . . . No gloves or masks. And one of them looked like a pretty good climbing tree."[48]

Environment in the Grocery Store

The trip to the grocery store gets similar treatment, although the strawberries in Ellen Richards's time would not have been available year round. Nor would they have been heavily sprayed with methyl bromide. In 2008, 2,708,710 pounds of methyl bromide were used on commercial strawberries in California alone. This despite the fact that methyl bromide reacts to create holes in the ozone layer that protects humans from sunburn and skin cancer; it was slated for phase out by 2005 under the 1992 Montreal Protocol to which the United States is a signatory. But the United States petitioned the Ozone Secretariat of the United Nations in behalf of California strawberry and Florida tomato growers, saying there was no good substitute for methyl bromide. Yet methyl bromide is also a strong neurotoxicant. Finally, in November 2010 a substitute was approved. Methyl iodide. It does not harm the ozone layer. It dissipates at ground level too quickly. But it is highly toxic, can mutate DNA, and is classed as a carcinogen. Workers who now apply methyl iodide to tomatoes or strawberries will need to wear respirators. As Steingraber, a worried mom and wordsmith puts it, "Given a choice, I'd like the strawberries and tomatoes I feed my children to be grown by people who do not require chemical weapons protection."[49] Steingraber wants us to understand, again, that being an alert parent or consumer is more than just a personal choice. It finally requires understanding and acting on the social forces that produce such dangers. She shops at an organic co-op, finding it more convenient, as well as healthy. But, she is quick to accompany this social medicine with a spoonful of sugary humor. Having forgotten to pick up shampoo at the co-op, she heads back there, warning Faith and Elijah that there will be no treats and no whining on their hurried stop. But little Elijah heads straight to the case with his new favorite snack—steamed kale with sesame seeds and tamari sauce. Told no one more time, he throws a toddler tantrum, falling to the floor crying, "Kale! Kale! I want kale!"[50]

Organic farming is not merely a lifestyle choice or simply a healthy one. It affects kids, workers, the entire planet. Similar journeys take us through the asbestos flooring in Steingraber's house, the dangers of PVC plastics, childhood asthma linked to school buses, the perils of natural gas fracking, and more. Along the way, she offers consumer tips, parenting reassurance, environmental information, and sound science served up in healthy, hearty family style. But it is more than ironic that in 2012, as she concludes *Raising Elijah*, she must once again give a call to action and join the fight to prevent fracking in the still-untouched deposits of the Marcellus Shale in upstate New York. Rachel Carson's Pennsylvania has been the center of the current natural gas fracking boom. Is all this polluting progress inevitable? Steingraber writes that she believes it is not. In this, she is indeed raising Elijah, not just setting an example for her own son, but invoking the name of the abolitionist martyr, Elijah Lovejoy. She describes, in a sad, alarming scene, how fracking has destroyed fields, farms, homes in Pennsylvania, and how it will permanently alter upstate New York, "the state's foodshed and wine-growing country" with some of "the largest unbroken forest canopy in the Northeast. . . . More than shale will be fractured." There will be well pads, roads, pipelines, the cutting of trees and the clearing of land through-out the fields and pastures of New York farmland. "Less obviously but no less real, fracking threatens migratory birds, some of whom are already suf-fering catastrophic declines." According to Audubon Pennsylvania, bird species threatened by fracking include Hermit Thrush, oven bird, veery, Winter Wren, and seven species of warblers, Scarlet Tanagers, and Wood Thrushes. Seventeen percent of the world's Scarlet Tanagers reproduce in the forests of Pennsylvania. Twelve percent of the world's wood thrushes lay their eggs there.[51]

Steingraber is following here in the tradition of Rachel Carson, Ellen Richards, and Alice Hamilton and their concerns with ecosystems, human health, and toxic chemicals. But she also combines and can claim direct descent from her upstate New York foremothers: Florence Merriam Bailey, the founder of amateur ornithology and the first student environmental organizer; Anna Botsford Comstock, the first female professor at Cornell, whose writing spread the nature-study movement; and Susan Fenimore Cooper, whose *Rural Hours* started the whole business of morally and polit-ically engaged writing about nature and how humans could destroy it. It is as if Steingraber has brought them all back to life, has read "A Lament for the Birds," or, perhaps, is simply channeling *Silent Spring*, with a newly updated, nonfiction version of "A Fable for Tomorrow."

6 ❧ RACHEL CARSON, DEVRA DAVIS, POLLUTION, AND PUBLIC POLICY

Like RACHEL CARSON, Devra Davis grew up near polluted Pittsburgh, got a degree from Johns Hopkins, and labored for years within the federal government. Davis is passionate, principled, and widely read. Her PhD is not in some highly specialized scientific subject, but in culture and science studies. Epidemiology and environmental health expertise came later. And like Carson, her principles and polished prose have drawn the ire of polluting industries and their ideological allies. Terry Tempest Williams and Sandra Steingraber are sensitive souls who seek to touch the spirit with their scientific subjects. Both have testified to Congress, both have joined in politics and protest. But their sensitivity and sense of place is not of Washington and its ways. Rachel Carson and Devra Davis both have cared deeply about chemicals and cancer; both have been familiar with the corridors of power. Interestingly, given her training at Hopkins and her concern with cancer and its environmental causes, Devra Davis also crosses paths literally, and in the library, with those whom Rachel Carson built upon—Raymond and Maud Pearl, Anna Baetjer, Alice Hamilton and Harriet Hardy, and one of Rachel Carson's most important sources on cancer, Wilhelm Hueper. A World War I veteran who became a pacifist, Hueper was smeared as a communist. Attacks similar to those against him were aimed at Hamilton, at

Carson, and at Davis herself. But by the latter part of the twentieth century and the beginning of the twenty-first, charges against environmentalism as some part of an un-American commie plot had morphed into supposedly "sloppy science" that would set back American industry and strangle our competitiveness through big government regulation.

No matter. Devra Davis is not as demure as Rachel Carson. At the end of her shortened life, Rachel Carson jabbed back at her critics. Davis throws roundhouse punches. Like Terry Tempest Williams and Sandra Steingraber, Davis is a feminist who artfully combines the personal and the political, powerful anecdotes, and accumulated data. But as the orphaned survivor of a family that has been haunted by the Holocaust and the horrors of pollution, Davis takes the offensive. Millions have died from modern industrial pollution, and more will, unless action is taken. Davis describes systematic cover-ups and foreknowledge of massive injury and death that she likens to crimes against humanity. Her work is a crash course, made creative, yet credible, in the history of environmental health and the human cost of industry's fight to prevent regulation or reform.

This strong stance evolved slowly for Devra Davis. Over the course of decades, she attempted to get the cancer establishment to face up to the role of environmental exposures. She sought, as a conscientious researcher and government official, to follow the science and the rules. But each victim of environmental pollution, disease, and death is a very real person to her; she draws on the ethos of the Holocaust. Victims of the Nazis or of pollution must never be seen as faceless numbers. The result is a powerful combination of modern science and morality, of caring for family and friends, and of emotional engagement that resembles the passionate religious and ethical stances, the circles of female friends and supporters, that mark earlier women writers.

Davis's work has included positions at the National Academy of Sciences (NAS) and at the University of Pittsburgh Cancer Institute, where she was head of its Center for Environmental Oncology. She has participated in countless government panels and studies, authored peer-reviewed and popular articles, and made speeches and media appearances. But her major contributions are captured eloquently in two well-received, gracefully written books for general audiences, *When Smoke Ran Like Water: Tales of Environmental Deception and the Battle against Pollution* (2002), a finalist for the National Book Award, and *The Secret History of the War on Cancer*

(2007). Davis recounts the history, dangers, and epidemiology of air pollution through strong narrative and compelling stories. A precocious child in a small town, Devra heard her grandmother and grandfather speak different languages—Russian, Hungarian, Romanian, Yiddish—and tell stories in them. She would hear and listen to stories on foreign-language radio stations and from nuns in school, from whom she learned to play piano and to love church music and the Bible stories that came with the teaching. "I became fascinated with stories and their sound and the idea that words and stories could have different meanings."[1]

A story is the starting point in *When Smoke Ran Like Water*. It is an environmental and personal history that argues for the precautionary principle—the same approach to the hazards of chemicals in modern life recommended by Rachel Carson. In the 1980s, Davis is working at the National Academy of Sciences. Senator Daniel Inouye (D-HI) has given a grant of a half million dollars to the Federal Aviation Administration (FAA) for the NAS to study the problem of why the senator, who travels regularly back to Hawaii, has been getting sick because of the flights. Davis is assigned to the case and carries a clunky dust monitor onto a plane. It measures fine particles and tells her immediately that the levels of smoke throughout the plane are the same as in the smoking section. No one is immune. She excitedly declares that there is no need for a major study and that money can be saved; a couple of more flights should do the trick. Inouye informs her that no one will accept such findings without major data, lengthy reviews, official imprimaturs. It takes four more years to gather such data. But even then, smoking is finally banned on planes not because of the overwhelming evidence that cigarette smoke is the culprit harming human health, but because Davis notices gunky yellow stuff on the bottom of the planes. The gunk has been fouling not only human lungs, but, more important to commerce, the engines and instruments of aircraft. The ban sails through.

Davis quickly brings us up to date by also recounting how in contemporary computer chip factories that feature "clean" rooms, it is, again, the computers, not workers, that are protected from chemicals and fumes. And, finally, she lets us know what else we are up against. When Rachel Carson revealed the dangers of pesticide manufacture, her publisher was threatened with lawsuits. Four decades later, Devra Davis tells how she is brought to a deposition hearing in a case against chip manufacturers—IBM,

FIGURE 6.1   Devra Davis, a leader on cancer and
its environmental causes. Photo courtesy of Devra
Davis the Environmental Health Trust Fund.

DuPont, Dow Chemical, and Hoechst Celanese. She is confronted by
eight men in suits who all greet her mockingly by saying "My name is Dean
Allison." They are reminding her that they can say whatever they want to,
even lie, because they are not under oath. She is the only one who is. She is
an expert in behalf of workers with cancer acquired through their exposure
in "clean" rooms to known carcinogens like benzene, asbestos, cadmium,
and solvents of many types. But the lawyers demand a smoking gun. As
in death by gunshot, unless the specific chemical, the specific exposure,
and its role in the death of a specific victim can be named, no one will be
liable.[2] Devra Davis wants us to know how such abominations have come
to be, why she writes and acts as she does, and, above all, what can and
must be done.

Statistics Are Accumulated Stories

Davis's approach to epidemiology is etched in her mind by the story of her friend Deb Filler's father, Sol. With his brother, Tootzie, Sol Filler survived the Nazi death camps. On August 2, 1942, in Brzozow, Poland, Jews were being shot in the streets. The mayor, a Nazi collaborator who was identifying Jews, comes to Sol Filler's house during Kiddush supper. The family had planned to flee that night. But the mayor demands that Sol's father, a baker, turn out four hundred loaves for the occupying troops, or he will shoot the family himself. Sol, his brother, and father are spared. Sol survives until August 1945 when the Nazis, fleeing in the face of the Red Army, force thousands of Jews to march five hundred kilometers (more than three hundred miles) through the cold to their former phony model camp, Theresienstadt. At the war's end, Sol finds himself at a displaced person's camp in Dachau. It is some forty years later before he can finally bear to visit the horrors of his youth. On a tour of Theresienstadt, a guide says that many Jews survived the march from Auschwitz to this place. "The Germans kept meticulous records on everyone who ever came here." Sol cannot remain silent. That is not true, he interrupts. "How could you say such a thing? Nobody was writing down the numbers. I came here from Auschwitz with only two hundred. We started with ninety thousand. We were eating grass along the way. No one was counting the bodies."[3]

The Holocaust was one of the great unmentionables in Devra's childhood, as was pollution. Davis carefully explains that she understands that the Holocaust, with its intentional, systemic killing and explicit plans, is not the same as the death of millions from pollution. These deaths are far harder to pin down, but they are real nonetheless. "The damages tied with our environment arise not from some deliberate effort to purify society against a presumed enemy from within but from the daily activities of people doing productive work." But the two evils do have something in common. Both are so grim and hard to grasp that they are not considered fit for normal, polite conversation. And there is the temptation to accept the consequences of pollution as necessary, to submit to the evils of denying or even lying about them, and to excuse those who would not have anyone count up the effects. "As a result, those felled by environmental conditions seldom even know why they are dying." And so, Davis, quoting the Talmud, which says that whoever saves a single life saves the whole world, has come to believe with Sol Filler that "it is a moral necessity that each fate be counted." We cannot change the past, but we can bear witness and try to change the future. For Davis it "makes no difference whether we are remembering millions, or a few felled by a hidden, local epidemic. Those of us who survive must enumerate, count, tally, and measure what has happened. We must record and pass on the truth."[4]

Carson and Davis Break the Silence

As with Sandra Steingraber's insistence that the title of *Silent Spring* underscores Rachel Carson's breaking the silence about the dangers of chemicals, of what has already been known but that remains unspoken, Devra Davis reveals why it is has taken so long to understand the dangers of toxic air pollution. She begins by talking about the silence of her own family members concerning Donora. In college at the University of Pittsburgh, Devra comes home and asks her mother about pollution in Donora. She has been studying the killer fog of 1948 in her own hometown; yet it is something she had never heard of before. Could there be another Donora, like there are other Allentowns, other Websters? She follows her mom into the kitchen. "I read in a book in school that in a town with the same name as ours, there was pollution. . . . I could not imagine that what I'd read had anything to do with where I'd grown up. I had never heard about our town being anything other than a wonderful place. I had never heard of pollution. The word sounded dirty, something to be ashamed of."[5] After much hesitation, her mother begins to describe the soot and grime and filth she had to constantly clean up. "Well, we used to say, 'That's not coal dust, that's gold dust.' As long as the mills were working, the town was in business. That's what kept your Zadde and your father employed. Nobody was going to ask if it made a few people ill. People had to eat."[6] Davis says of this conversation with her mother, "So Donora was famous, but no one ever talked about it. We lapsed into silence."[7] Florence Merriam Bailey on the slaughter of birds, Alice Hamilton on the dangerous trades, Rachel Carson on the perils of DDT, Terry Tempest Williams on nuclear fallout, Sandra Steingraber on groundwater that gives you cancer, Devra Davis on death from polluted air. To remember and name the victims, to break the long-standing silence, each needs a *Silent Spring*.

Devra tells the story of Donora's killer fog in intimate detail, with names, people, places. When at last the fog is over, the town's medical examiner says people should just go back to work. There is no exact count of the deaths, though the town has run out of coffins. Such a catastrophe is an act of nature and unlikely to occur again. The head of the Public Health Service refuses to investigate, calling Donora "a one-time atmospheric freak." But Davis reminds us that this is not a freak at all. Similar conditions related to fluoride from zinc smelting had already been recorded along the Meuse River in Belgium in the 1930s. Six years after that fatal disaster, researchers had calculated and warned that an even larger disaster could happen under similar conditions in London, where they estimated some four thousand people might die. As it turns out, there also had been warnings in Donora and throughout the Monongahela Valley. The Steelworkers Union had suggested capturing the pollution and reusing materials from it. No one listened. A major health study was begun but never completed; its records were never found. Officials up and down the line denied the dangers and avoided calls for accountability. It would take another killer fog—with more than four thousand deaths—in London in 1952 and decades of struggle by activists, epidemiologists, and a few concerned politicians to break the pattern. Often, the answers are hidden in plain sight. Davis reexamined many records from the Donora disaster years later. In the files, she found some old, simple maps and made some calculations. The answers had been there all along. As had been the case along the Meuse, and simply denied within Donora, the bulk of deaths were in homes that surrounded the town's zinc smelter or that matched the plume of pollution and fluoride gas pouring from it.[8] Thousands more, unnamed, uncounted Donorans would die over the years; untold more would suffer from related diseases before they died. Devra's beloved Bubbe, her Uncle Len, and many others suffered heart attacks and other ailments before they died—their deaths not numbered, never linked precisely to the smelter in the town. Davis wants us to remember.

The Struggle against Smog

And so we are introduced to a swift, interesting history of air pollution, especially from coal, which was first banned in England in 1306 by Edward I. Its smoke and fumes have been considered noxious throughout the centuries. Davis will not accept the "we could not have known" explanation for pollution deaths. In the 1980s, she and a colleague carried out a retrospective study of the thousands of deaths from the now notorious London killer fog of 1952. As with Donora, when the fog passes, officials breathe relief and declare all is well. But London citizens continue to die for many months. The official explanation is that it must be from the flu. Any other explanation would mean an end to coal burning or a turn to more expensive, efficiently burning coal that is sold for export. Davis and Michelle Bell, a graduate student from Johns Hopkins, set out to estimate air quality (average sulphur dioxide) and average death rates in London, before and after the fatal smog. They gathered detailed, but scattered, records and statistics kept by the British. Then they made a graph that traced both the dirty air and deaths in 1952 and 1953. What they learned is that adverse health effects and dirty air persisted in London even when the worst was over. As a result, compared with carefully kept historic averages, more than thirteen thousand people died. The conclusion was as clear as it was chilling. The deaths noticed at the time after the fog were not the result of flu and were far more numerous than was presumed. And the killer was coal.[9]

But how much air pollution does it take to kill and how do we know? The answers to these questions are still being battled in the Congress and the courts. But it is not for lack of science. The answers are embattled because of the stubborn resistance of polluters to the growing science of environmental epidemiology. For Davis, the personal is political; she begins with Uncle Len. An early fitness buff and runner, Len and his wife left dirty and declining Donora and, like many Americans, drove west to settle in the presumably cleaner, fresher air of the fabled city of Los Angeles. By 1955, however, General Motors had already bought up and eliminated trolley lines nationwide, including the handy Red Line trolleys that ran throughout the city. U.S. Senate Counsel Bradford Snell noted in a documented 1974 inquiry that as early as 1921, American car sales had gone flat; major cities were served by 1,200 electric intercity railways and street car lines. There were forty-four thousand miles of track and three hundred thousand employees serving some 15 billion passenger trips, which generated more than a billion dollars in revenue each year. Nine out of every ten vehicle trips were by streetcar. Then, in 1922, Alfred P. Sloan of General Motors created a special unit charged with eliminating street cars and other electric rail. Within a few decades, a consortium of GM, Goodyear, Phillips Petroleum, Firestone

Tire and Rubber, Mack Truck, and others set up shell companies in major cities that bought up rail lines and replaced them with buses and cars. These companies had been investigated and found guilty in 1947 during the Truman administration under the Sherman Antitrust Act. But it already was too late. The penalty? A fine of $5,000 or, as Davis puts it, the price of two small cars.[10] Southern California grew, highways were built apace, and Uncle Len and his wife, Aunt Irene, moved in. By 1955, there were five million people living in the Los Angeles basin, which is surrounded by hills and small mountains that trap polluted air. Smog was born. When Uncle Len, the fitness buff, succumbed to a heart attack, it was said it ran in the family. As Davis explains, it surely does. So does the exposure to pollution—ozone and particulates that bring on cardiovascular disease. What we have learned from Uncle Len's demise is that it is not only huge doses of badly polluted air, as in London or Donora, that bring on disease and death. Modern science now shows that long-term, lower dose exposures, as in Los Angeles, can be just as deadly.

As we learned in chapter 2, Devra Davis knew and admired the work of Mary O. Amdur, who undertook studies after Donora on the effects of sulphur dioxide exposure on human beings. Amdur invented special monitoring devices, revolutionized how we measure the effects on laboratory animals, and first showed that even tiny exposures could hurt both guinea pigs and people. Meanwhile, as smog and illness worsened in California, Governor Earl Warren, and later Governor Pat Brown, passed laws to limit air pollution and created studies to find just what the health effects of smog might be. In those days, smog was believed to cause mainly irritation of the lungs and eyes, not chronic disease and death. But, among other efforts, John Goldsmith published a study in *Science* magazine that showed that the particulates in smog—they formed around the sulphur dioxide that Mary Amdur studied—could cause cardiovascular disease. As Devra Davis puts it, that is how my Uncle Len "made it into *Science* magazine, about two years after his last handball game. His death became part of what scientists call *data*." The report by Goldsmith and his colleague, Alfred Hexter, was called "Carbon Monoxide: Association of Community Air Pollution with Mortality," and appeared on April 16, 1971. It showed that heart attacks and other deaths were highest on the days just after high levels of carbon monoxide had been measured in the outdoor air. "Within their analysis, buried beyond recognition, is my Uncle Len."[11]

But the early studies and legislation in California set the stage for the Clean Air Act of 1970 and nationwide air quality standards based on the best science, not cost or convenience. But even in this early heyday of environmental concern, opposition from industry was intense. William Ruckelshaus, the first administrator of the Environmental Protection Agency, recalled that he was told that setting the new air quality standards would be simple. There was a broad consensus; all parties were aboard. All, that is, except the auto industry. When they heard about the proposed new standards, they howled, fought, and sought delay. Next came a struggle over leaded gas, which had been an issue since Alice Hamilton's day. The Ethyl Corporation was still fighting in the 1970s and got delays in court. The last cars to run solely on leaded gas were manufactured in 1976. But it was not until 1995 that leaded gas, a known health hazard, finally was gone.

Smog and Uncle Len's death in Los Angles, early environmentalism and the regulations in California were precursors of things to come. Savvy politicians like California's Richard Nixon put their fingers in the particle-filled air and decided to ride the new winds. Given his concern that the environmentalist Maine Senator Edward Muskie would be his likely 1972 opponent and with the outpouring of massive, peaceful environmental demonstrations across the country on Earth Day in 1970 (Washington, DC, was merely the biggest), Nixon became a surprising supporter of environmental regulation. He created the EPA, even while privately making vicious fun of environmentalists as "a bunch of Commie pinko queers!"[12] But the EPA was immediately controversial and underfunded, and it was countered by industry, especially automakers, with complaints, stalling tactics, and counterattacks.

How Industry Creates Confusion and Fights Clean Air

Devra Davis explains how the game of doubt, denial, and delay is played. In order to improve air quality, autos needed to be improved. But health studies always tend not to be black and white or definitive. There is room for quibbling and doubt, which industry has mastered. Early studies on environment and health were often challenged, ignored, or denied by the public health establishment itself. National health statistics only truly emerged after World War II and were improved when states needed to gather more information about the health of their citizens in order to meet the requirements of Medicaid and later Medicare. Such improved information allowed larger studies of national trends. After the surgeon general's report on the health dangers of smoking, similar warnings about air pollution began to emerge. In the most polluted areas of Buffalo, New York, it was discovered that residents had more bronchitis than elsewhere. And, in the early 1960s, the *Journal of the National Cancer Institute* reported that both smokers and nonsmokers in urban areas (where the air was more polluted) had a higher risk of lung cancer death than rural Americans. By 1970, two young researchers, Lester Lave of Carnegie Mellon University and his undergraduate assistant, Eugene Seskin, set out to create a more definitive approach to the health effects of air pollution. The two quantified the health costs of those affected by bad air in 114 metropolitan areas where sulfur oxides and particulates were now being measured along with the deaths of infants and residents over sixty-five. They used statistical analysis to account for variables such as income or smoking. Then they put a price on the positive contributions to the Gross National Product (GNP) that would have come from those who died and created the concept of a public good. The productivity of those who were killed by air pollution adds up to more than the "losses" to the GNP from the unproductive hospital stays, burial costs, and the like. As Devra Davis, puts it, "green accounting" was born.[13] Other studies began to reveal patterns of the harm from air pollution, especially one called CHESS (Community Health and Environmental Surveillance System) carried out by Jack Finklea, a young doctor at the EPA, who would later head NIOSH (the National Institute for Occupational Safety and Health). Both early studies clearly indicated the dangers of polluted air to the health of American citizens. Both were attacked.

Lester Lave was an economist, not a physician, and at the time even doctors were still skeptical that an individual whose lungs did not reveal classic damage might still be suffering harm. The concept was simply too new, too challenging to orthodox ideas. These circumstances left Lave and Seskin's study vulnerable to attack by researchers paid specifically by industry to challenge every piece of evidence, methodology, or doubt. In this case, the industry group representing coal-fired utilities, the Electric Power Research Institute, hired Ron Wyzga of the Harvard School of Public Health to do just that.

Jack Finklea fared no better. Within the Nixon administration, his work documenting disease and death from air pollution in major cities reached the attention of the Office of Management and Budget when he requested more funds in 1972. James Tozzi of OMB told him directly that more detail was *not* wanted. "Dr. Finklea . . . I don't doubt you could use more money to get better information. In fact, that's precisely the problem. I think the Congress does not need to hear all these details. I don't think the country is ready for that." Funding was cut, CHESS ended, and no final report ever appeared that would have revealed lethal smogs in New York, Birmingham, Pittsburgh, and many other industrial centers that had led to hundreds and hundreds of deaths, including many throughout the Northeast over the Thanksgiving weekend of 1966.[14]

Then, in 1980, John Spengler of the Harvard School of Public Health teamed up with his mentor, the physicist Richard Wilson. They validated, improved, and went beyond earlier studies by Lave and others. With new research from the Brookhaven National labs and elsewhere, using stronger tools and analyses of wind, weather, and descriptions of the engineering characteristics of electric power plants, along with detailed emission and health data, they estimated that in the year 1960 alone, fifty thousand people in the United States had died from air pollution. Their book, *Health Effects of Fossil Fuel Burning*, made headlines.[15] But as Spengler later told Devra Davis, "One of my medical colleagues—far senior to me—dropped by. He looked at my new book and asked, 'Mind if I borrow a copy? I'm being paid to criticize it.' That was my introduction to how the game was played."[16] The tactics had been set for the long-running fight against healthier air standards that continues to this day.

A long battle ensued during the Clinton administration when EPA Administrator Carol Browner simply tried to update the National Ambient Air Quality Standards (NAAQS), as was required by the Clean Air Act. Industry again tried to attack the research and researchers. Among many pro-pollution lobbying efforts, the so-called Center for Regulatory Effectiveness was chaired by none other than James Tozzi, the OMB official who, in 1972, had denied more funding for air pollution studies.[17] Then, for almost a decade, no progress at all was made on air quality standards during the George W. Bush administration.

There is a profound human cost to all this denial and delay. As Davis says, it has long been known, and then firmly established, that polluted air kills more than sixty thousand Americans a year and that breathing polluted air is the equivalent, in terms of lung cancer risk, of living with a heavy smoker. For every 10 microgram increase of particulate pollution in the air, there is an 8 percent increase in the rate of death from lung cancer.[18] For Davis, these are not simply health statistics. "If the technologies readily available in 1980—nothing exotic, just scrubbers and electrostatic precipitators on power plants, better use of wasted heat and engine exhausts, and the use of more efficient lights and energy—had been in place to reduce emissions from power plants and transport, starting that year, well over a million people would have been spared earlier deaths. How much expense, how many missed quarterly profit projections, how many inconvenienced or even downright angry lobbyists are a million lives worth?"[19]

Once they have been labeled "controversial" or "scientifically suspect," no matter how loony the source of the accusation in this Internet age with its conservative media echo chamber, many researchers, writers, and advocates would be deterred. But Devra Davis soon was in the midst of the world's biggest and most controversial environmental health challenge—that of global warming. In an important *Lancet* article in 1997 on air pollution, dealing with both science and policy, Davis and colleagues linked air temperatures, the health effects of bad air quality, and the rising temperatures associated with global warming. The result was staggering. Davis and her team of experts had calculated that eight million deaths could be avoided if modest greenhouse gas reductions recommended by the European Union were implemented by 2020. The article caused a minor sensation and was immediately attacked by defenders of industry. But Davis was undeterred. I worked the media alongside her at the Kyoto Protocol in the cavernous, cacophonous conference center that held delegates, NGOs, lobbyists, and the press—some sixty thousand people in all. Her advocacy was exactly the right role for scientists who had become aware of the dangers to humanity from the mindless burning of fossil fuels. Few scientists could have effectively combined research that was accepted by a leading, peer-reviewed medical journal, as did Devra Davis, with the cheery, cheeky chutzpah and speedy sound bites required to approach and lobby crowds of exhausted delegates and jaundiced journalists from around the world.[20]

Such progress as has been made in environmental health studies has not simply been one of disinterested science versus industry. Rachel Carson was, of course, concerned with the connection between chemicals and cancer. But she shied away from connecting her own cancer to chemicals, to discussing it in public, and, though she was clearly deeply politically engaged and worked tirelessly with women until she died in 1964, did not consider herself a feminist. A generation later, after the rise of feminism and leaders like Betty Friedan, Gloria Steinem, and Bella Abzug, Devra Davis, like Sandra Steingraber, is unwilling to hold back, to let women die, or to ignore the environmental causes of cancer—especially breast cancer. Davis wants the public to understand that it has taken the efforts of organized citizens, especially women, to advance the cause of environment and health after Rachel Carson's time. Like Sandra Steingraber, Davis recalls marching with women against breast cancer who carried signs reading "Remember Silent Spring." Davis still has a button from that march, in pink and black, that reads "Remember Rachel Carson." She is convinced that protesting women in the breast cancer movement "rekindled an interest in Rachel Carson that had not been there for years. That's what caused me to go back and re-read her." Devra Davis is proudly part of that broader women's movement.[21]

The Influence of the Women's Movement on Breast Cancer Research

Devra Davis did not meet Bella Abzug, the legendary New York congresswoman and antiwar, feminist, and environmental leader, until the 1990s. She received an unexpected phone call. The booming, gravelly, familiar New York voice said, "This is Bella. My friends are dropping like flies. I'm gonna have a hearing at City Hall on breast cancer and the environment. March 20. I want you to be there." Abzug did not have cancer but was horrified by its proliferation. When she was a child, only one in forty American women developed breast cancer. "By the time she was student body president at Hunter College and editor in chief of the *Law Review* at Columbia University the rate was one in twenty. Now in her seventies, Bella wanted to know why breast cancer attacked one in nine women by age eighty-five."[22] She particularly wanted to know why breast cancer rates were four times as high on Long Island as in other areas and believed that if the link to some environmental contaminant could be shown, as had been done with nuclear fallout and strontium-90, the battle could be won.

Davis explains that the first statewide cancer registry was instituted in Connecticut in 1941. By the 1990s, there were many such registries, though no truly nationwide database. But national maps of the prevalence of the disease could and had been made. They showed that the highest rates for the "sisterhood of breast cancer" were in the Northeast, the Great Lakes states, and California. They were lowest in the South. Breast cancer is highest near areas with more industrial and urban pollution. By 1998, breast cancer was the most common cancer among women in the world. The average American woman who dies of breast cancer has her life shortened by about twenty years. In 2001, there were forty-two thousand deaths annually from breast cancer. Although the death rate has declined in recent years, owing to early detection and treatment, the incidence continues to rise. It was why Bella Abzug wanted hearings and how Devra Davis got involved. There were hearings, marches, media, and ads. One featured the top fashion model Matuschka, really Joanne Matuschka from New Jersey, who had lost a breast to cancer. She posed for a then shocking ad to reveal her loss and the harm to women as a result of ignoring the disease. While hiking together at the First World Congress on Breast Cancer, Matushka explained to Davis why she had become an activist, had posed, and linked all this to her childhood in the days of *Silent Spring*. "I grew up in New Jersey, spent summers at the shore, with lots of pesticide spraying all the time. Everybody in my old neighborhood is getting the disease now. . . . I can remember my grandfather keeping gallons of stuff around the house in the garage and all those trucks spraying mists that we would run behind, pretending we were in the Twilight Zone."[23]

The links between Rachel Carson's warnings and Sandra Steingraber's work are, for Devra Davis, quite explicit. "Rachel Carson made it clear why DDT should never have been used as a mainstay of agriculture. After Carson, it was no longer possible to measure progress in terms of the volume of synthetic chemicals applied to cropland or the number of insects killed." DDT-resistant mosquitoes evolve quickly, while beneficial birds and insects are killed. And, Davis explains, DDT and its metabolites, or slightly altered residues, accumulate in animals with fatty tissues, affecting their hormones. "As Sandra Steingraber reminds us in *Having Faith*, her poetic exploration of pregnancy and childbirth, at the top of the human food chain lies the nursing infant, taking in her mother's milk and everything that comes with it."[24] But pinning down which chemicals are guilty and how they promote cancer is difficult. Davis explains that, in addition to the metabolites of DDT, there are a number of persistent chemicals that may affect hormones. Other chemicals that cause mammary tumors in rodents—what would be breast cancer in humans—leave no traces in the human body only minutes after exposure. Such compounds, including benzene and butadiene in gasoline and car exhaust, are widespread, as are phthalates and other ingredients in plastics, and more than forty other chemicals already tested by the U.S. National Toxicology Program.[25]

Soon, grassroots breast cancer coalitions led by women were springing up. It is here that Davis's contribution—serving as a link between the scientific, medical, and environmental health communities, reaching out to a broader public, and promoting the role of serious activism—shines. Though Davis is clearly extroverted and willing to talk about herself—her inclusiveness and willingness to give credit to others is admirable. She draws, for instance, on the work of Breast Cancer Action (BCA) leader Barbara Brenner to show that there are many, many women, leaders, and organizations involved; and they are just the most visible parts of the women's breast cancer movement. Among many activists whom Davis credits are Karen Miller of the Huntington Breast Cancer Coalition and Lisa Bean, a social marketer of the Women's Community Cancer Project. Bean designed colorful protest signs for an original massive protest march on Mother's Day in Boston in the early '90s and then produced stirring photos of the more than four thousand women who marched. Additional activists and leaders were represented at Bella Abzug's hearings. Davis describes how Elinor Pred, who founded BCA in San Francisco in 1990, took a page from AIDS activists in starting one of the first groups to address the environment and women's health. BCA held a creative and unusual event in 1992 at the historic Palace of the Fine Arts in San Francisco with Susan Love the activist breast cancer surgeon, Davis, and others. Davis was flabbergasted by the strong response. She also "got a look at what women could do if they organized themselves."[26]

Such women activists traced clusters of disease on Long Island and elsewhere, demanding a response. Soon, a few powerful male politicians who controlled funds, like Senator Alphonse D'Amato of New York and Senator Tom Harkin of Iowa, responded. By 1993, Andrea Martin and others, Davis tells us, had pulled the many breast cancer groups together into a major coalition that included NOW, the Susan Komen Foundation, and others. It successfully pushed for increased funding for research into the links between the environment and breast cancer. By then Bella Abzug had been diagnosed and was dying of breast cancer herself. Her portrait, along with other notable women who had died of the disease, like Rachel Carson and poet Audrey Lorde, was on a colorful community mural painted by Be Sargent on the wall of the popular Loew's Theater in Cambridge, Massachusetts, where much of the action had begun.[27] Davis notes that she was humbled, but proud, to have even a small portrait of herself in a panel off to the side, to be included with her heroes who had gone before and died. By 2002, the Breast Cancer Fund and Breast Cancer Action had pulled together scientific studies from around the world that showed the connections between synthetic chemicals and breast cancer. Their white paper showed the connection between phthalates and organochlorines—both widely used—and the likelihood of cancer. "We ignore at our peril the increasing evidence that chemicals are contributing to the rising tide of breast cancer."[28]

Chemicals, Cancer, and Men

Devra Davis was also involved with one of the most amusing or scary studies of the environment and human health that caught the attention of male policymakers. In the 1990s, more and more evidence was emerging, as we shall see when looking at Theo Colborn in chapter 7, that even small doses of certain toxic chemicals were disrupting human hormone production and affecting human reproductive health in both men and women. These ideas drew on the work of researchers like Louis Gillette at the University of Florida, who showed that alligators that had been exposed to pesticides in Lake Apopka in Florida were growing up with smaller penises. Other studies showed that in human males exposed to toxins, sperm counts were dropping drastically, sperm motility was reduced, and fertility was impaired. Other serious effects in males include testicular cancer, a higher rate of stillborn boys than girls in Japan, and a declining ratio of boys to girls among infants who are born in the United States. It was Devra Davis who, with a colleague, pulled together a variety of studies that indicate declining sex ratios for the birth of boys.

All this activity, as well as the attacks on passive smoking and other environmental health campaigns, once again led to a reaction from industry, this time led by the powerful tobacco group. They sought to reinforce the use of risk assessment (a methodology that presumes some risks are necessary and tolerable and that they can be prioritized by the cost to reduce them). Industry then further worked to downplay the risk ratios that should be considered worrisome. Manufacturers formed a coalition called the Advancement of Sound Science Coalition (TASSC). Unknown—except to its founders, when it was formed in 1993—was the fact that the tobacco company Philip Morris had put up the first $320,000 and that the PR firm that handled TASSC had already been deeply involved in attacks on environmental health science. TASSC then sought legitimation by including less obviously biased product groups like the Chemical Manufacturers Association (CMA), even though Philip Morris's Ellen Merlo, vice president for corporate affairs, wrote to CMA explaining that the overriding objective of TASSC was to discredit the EPA report on smoking and to get EPA to adopt a low risk assessment standard for *all* the products they review.

Only later was it revealed that the brains behind the Philip Morris front was, again, none other than James Tozzi, the OMB official who originally killed air pollution funding under Richard Nixon. He was paid a retainer of $40,000 a month during 1993 and $600,000 for the year 1994. By the time Davis wrote her 1998 article for *JAMA* on sex ratios, TASSC had dissolved in all but name, but its mantle and website, called "Junk Science," was carried forward by Steve Milloy, a journalist and lawyer. When Davis's article appeared in the peer-reviewed, prestigious *JAMA*, it drew on the work of noted researchers around the world. Nonetheless, Steve Milloy and "Junk Science" went on the attack. "The amazing and incredible Devra Davis is at it again. . . . Davis now claims that the proportion of males born has declined and that this can 'be viewed as a sentinel event that may be linked to [man-made chemicals in the environment, so-called "endocrine disruptors" or "environmental estrogens"].['] But Davis' claim is not based on a scientific study. This article is merely a collection of anecdotes woven together to cause alarm."[29] Devra Davis could point out that she was in proud company. Steve Milloy and "Junk Science," it would turn out, were among the first to renew the attacks on Rachel Carson, long after she was dead, as the original proponent of "junk science," supposedly responsible for the death of Africans because of her exposés of the dangers of DDT.

Devra Davis and the Secret History of the War on Cancer

In 1986, with PhD and MPH in hand, Davis was already an accomplished, rising environmental star working at the National Academy of Sciences under Frank Press, an MIT professor and former science adviser to President Jimmy Carter. She and her colleagues, with prestigious grants, had been studying and writing in journals about the fact that various cancers were still increasing and that the rise could not all be explained by tobacco, aging, or improved diagnosis, as was then the standard explanation. She now proudly comes to Press to tell her mentor that she has been offered a book contract to write about her findings that the nearly two-decade-long war on cancer is not going very well. Press looks at her and in his seasoned, diplomatic style says, simply, "It had better be a good book." Davis gushes how much the publisher likes it and that she has been offered far more in royalties than a typical first-time author. He repeats, "It had better be a really, really good book." Naively, Davis replies that her publisher expects it to be good. She doesn't understand. Press explains, "It had better be, because you won't be able to work here after you write it. . . . I'm just telling you that you can't write a book critical of the cancer enterprise and hold a senior position at this institution."[30] Another friendly man, temporarily heading the National Institutes of Health (NIH), offers similar advice. He finds her work on tracing cancer patterns quite interesting. "You know, I started out my career interested in the environment and cancer. I'm pretty sure that some of the lung cancer we see in women in southwestern Pennsylvania, where you come from, has something to do with the environment. I actually tried to do a study on that when I started out doing research, but I decided against it." Davis asks him why. "You ever hear of Wilhelm Hueper?" At that time, she had not. Her friend explains that Hueper was full of good ideas on the environment and cancer and believed that an exclusive focus on tobacco would take away concerns from causes equally deadly. But he was railroaded out. "Somebody like you should think about that."[31]

Devra Davis does think about it. And in the middle of the Reagan administration, with steady assaults on government regulation and the environment itself, the timing does not seem right to publish a book challenging the cancer establishment. So for the next ten years, Devra Davis turns out solid research papers, referenced NAS reports, and more that simply examine the relationship between the environment and health—of chlorinated water, smoking, and other subjects. Each ends with a familiar conclusion, almost a refrain. "More research is needed."

Then, in the 1990s, Davis uncovers several important archival sources, never before used. The files reveal that researchers have known about the environmental causes of cancer even far longer than she suspected and that the information has been covered up. The records came to her attention while continuing research for her long-postponed book. They take us back to the 1930s, to a forgotten international congress on the causes of cancer, providing important evidence about why the knowledge of environmental causes of cancer has been suppressed—and introducing two colorful, compelling characters, Wilhelm Hueper and Captain Robert Kehoe.

Before Rachel Carson: A Generation of Lost Research

Davis begins this story with a dramatic, reconstructed voyage and journey taken over four weeks in 1936 by the University of Chicago pathologist Maud Slye. Slye, also now forgotten, was considered the Madame Curie of her time. *Time* magazine reported that this celebrated woman, who had been able to breed rats with and without cancer, was headed to Brussels, Belgium. There she met up with more than two hundred of the world's top cancer researchers to review the latest findings about this dread disease. Her curiosity piqued by this brief *Time* reference, Devra Davis scours medical libraries, puts a talented reference librarian on the trail, all to find a copy of the actual conference presentations. Finally, an old three-volume set of the papers arrives from a library in Belgium. It is the proceedings of the 1936 Second International Congress of Scientific and Social Campaign against Cancer.

Davis had expected ignorance, outdated theories, naive science. She was wrong. "I spent a sleepless night fascinated by the sophisticated drawings and advanced research techniques that were employed to unravel the causes of cancer before I was born."[32] William Cramer of Britain had outlined how cancer was now one-third more common than at the beginning of the twentieth century and the reason was not, as had been said, because of aging and better diagnostic techniques. He also documented, with studies of identical twins, that cancers depended on where they lived and worked, not their shared genes. Cancer is not inherited. Cancer was the result of past exposures, sometimes twenty years or more. What was needed were experiments on animals who reproduce much faster than humans to better understand the causes. Angel Roffo of Argentina demonstrated that exposure to invisible radiation like ultraviolet and X-rays could cause cancerous tumors in rats and thus in humans. Other numerous and comprehensive clinical and laboratory reports clearly linked cancer to a variety of commonly used substances like arsenic, benzene, asbestos, synthetic dyes and hormones, and solar radiation. One German study offered a handy chart on the frequency of cancer deaths in Bavarian men by occupation, with professionals at the top, with the lowest rate; through skilled and unskilled workers; and farmers, at the bottom, where the rates were far, far higher. Davis had expected "to find amusing errors." Instead, "The papers did not depict the dark ages of cancer research but rather an exhilarating time of lively and important work that seems to have come and gone like a comet."[33] Roffo had publically warned of the skin cancer dangers of tanning and sunbathing. J. W. Cook and Edmund L. Kennaway reported on more than thirty studies in England that showed regular exposure to estrogen produced mammary tumors in male rodents. Yet, in our time, Davis reminds us, the National Toxicology Program of the U.S. government "did not formally list both estrogen and ultraviolet (sun) light as definite causes of cancer until 2002."[34]

Davis also reviews the gruesome role and research of German Nazi doctors, not to defend them, but to show that following Hitler's quest for a superior and healthy race, they researched and led campaigns against tobacco and cancer, synthetic cancer-causing dyes and other products, and promoted natural, organic diets. It is against this background that Devra introduces the two men interested in German environmental and occupational medicine who turned out to use them in far different ways. One is Wilhelm Hueper, whose findings Rachel Carson cited, and whose path Davis was warned not to follow. The other is Captain Robert Kehoe, a U.S. Army field investigator for the Office of Strategic Services (OSS), the predecessor of the CIA, whose files show that the doctors who helped promote racist theories, forced eugenics, and carried out human experiments were in many ways ordinary professionals who went along or were swept up in what seemed "normal." After the war, German scientists and doctors were hanged as war criminals along with military leaders. In an American context, the postwar stories of our two leading researchers are deeply instructive. Wilhelm Hueper tried to use German science to prevent environmental cancers, while Robert Kehoe kept their helpful findings secret in order to promote himself and to aid American manufacturers.

Wilhelm Hueper, Rachel Carson's Source

Hueper, one of the main sources for Rachel Carson's breakthrough writing on the environment and human health, was born in Germany and served in World War I, but became a pacifist and progressive. He emigrated to the United States as a married physician, a pathologist who ended up at the cancer research laboratory at the University of Pennsylvania Medical School in 1930. At the time, it was heavily supported financially by Irénée du Pont, head of the DuPont Corporation. Impressed by what he thought was du Pont's open attitude, he wrote him in 1932 that German research had shown that men producing synthetic dyes similar to those at DuPont were showing signs of tumors in their bladders. Hueper warned that there would be many cases at the dye plants in Delaware in the future. He was told there were no bladder cancers at the plant. "Well, that may be," he said, "but they will get them." In 1933, after layoffs and unemployment in the United States, Hueper went back to Germany, where Hitler had just come to power. Hueper still could not find a job, but he was appalled at what he saw. The Nazis had banned experiments on dogs and rabbits, but carried them out on humans—Gypsies, homosexuals, and Jews. Hueper returned to the United States, where twenty-three cases of bladder cancer had turned up at DuPont. In 1934, he was hired as the pathologist at the newly created Haskell Laboratory of Industrial Toxicology in Wilmington. Hueper was now familiar with medical literature going back to 1895, in Germany and elsewhere, that showed the connection between dye production, including two of the leading ones that DuPont was making, and carcinogenesis. He began a system of monitoring workers, but after being taken to a cleaned-up facility (like Alice Hamilton two decades before) he dropped in unannounced at a different one; he found hazardous dust wherever he looked. Hueper would never set foot in a DuPont facility again. He was ordered to confine his research to the labs. In 1937, he was ordered not to publish any of his findings. He protested and by 1938 was fired.[35]

Hueper pulled together his research and experience and did publish his major work in 1941. Called *Occupational Tumors and Allied Diseases*, it compiled more than a century's epidemiologic and experimental work from four different continents; it revealed that workplace factors were critical and controllable factors in cancer and other illnesses. As Devra Davis says, "The book was intended as a public health call to arms."[36] But like Rachel Carson's first, highly regarded book, *The Edge of the Sea*, also published in

1941, the advent of World War II soon obscured these key works. Hueper did write important articles that were well received in the 1940s in *JAMA* and elsewhere on the cancer risks of aromatic amines, estrogens, coal tar, arsenic, asbestos, and other hazards. Finally, he was named head of the U.S. National Cancer Institute (NCI) in 1948. There he led its first ever section on environmental cancer, provided original research, and synthesized the available world literature on the avoidable causes of cancer. By 1950, while Rachel Carson was also producing government pamphlets for the U.S. Fish and Wildlife Service, Hueper supervised a pamphlet from NCI designed for the general public. It was blunt in warning about the environmentally induced, avoidable causes of cancer.

It said that cancer had been known for centuries, but that modern life was shifting the patterns. Medical X-rays, dietary deficiencies, drinking, tobacco, sunlight, and toxic chemicals all played roles. In fact, Davis reports, the twenty-page brochure, complete with illustrations, "fingered workplace causes of cancer raging from radiation to specific toxic chemicals such as asbestos, aniline dyes, aromatic amines, paraffin oil, shale oil, crude oils, benzene, chromates, and nickel carbonyl." Risks did not stop at the factory gates either. Residents near such factories had higher cancer rates, too. Later studies would reveal, Davis says, that Salem County in Delaware, the location of the DuPont Chambers Works that Hueper studied, had the highest rates of bladder cancer in the nation.[37] But soon, we know, both Hueper and Carson would be attacked for showing that chemicals cause cancer. When Hueper sought to publish an update to his environmental health pamphlet in 1959, as Rachel Carson was writing *Silent Spring* and struggling with breast cancer, the NCI editorial board delayed so long that it never appeared. Later, he even was unable to get permission from the NCI to update his monumental *Occupational Tumors and Allied Diseases*.[38] Ultimately, Hueper was no longer allowed to speak to medical students and became seen as antibusiness and procommunist. He then discovered that his publications were being supplied by someone in Washington to the management of the DuPont Corporation and on to the Haskell Laboratory, where he had worked, for review and rebuttal. Hueper also ran afoul of the government when he prepared and submitted a manuscript on the cancer dangers of mining chromate ore and uranium, two essential ingredients of nuclear weapons. He was told that the Atomic Energy Commission had objected and there are other reasons why uranium miners develop cancer, not radiation. He must omit it. Hueper sent his findings elsewhere and soon received a letter from the Federal Loyalty Commission telling him he was under investigation. A former boss at DuPont had written that Hueper showed communistic tendencies. Once again, Hueper was forbidden to do further research on workplace cancers and told to stick to rats. He hung on at NCI until 1968, but was kept from publishing under threat of legal action.[39] Wilhelm Hueper ended his career as an alienated outsider. But the story is far different for another man who knew German science and was an expert on the dangers of the workplace—Captain Robert Kehoe.

Robert Kehoe Helps Hide Environmental Health Research

Robert Kehoe graduated from medical school at the University of Cincinnati in 1920 at the time that Wilhelm Hueper was completing his medical degree in Germany. His experience from the beginning of a steadily rising career showed the influence of industry. Leaded gasoline was introduced in 1923. But after two plant workers, who bottled liquid lead for General Motors in a small facility in Dayton, Ohio, died, the production line was temporarily shut down. Charles Kettering, the leader of GM efforts to develop leaded fuel, blamed the workers. Kehoe, a young assistant professor of pathology, was brought in to advise. He said that the danger was in the bottom of the plant, where fumes accumulated. Fans and hip boots and training for careless workers should do the trick. When it became clear that working with lead was literally driving men insane, Kehoe still assured the company that creating less sloppy procedures would eliminate the problem. But Standard Oil and the Ethyl Corporation, the very corporations that Alice Hamilton battled, went to great lengths to keep industrial fatalities secret. One worker, Joseph Leslie, made liquid lead at the Bayway, New Jersey, plant. After a 1932 explosion of liquid lead he was locked away in a psychiatric hospital. It was said that he had died. Only his wife and son knew he was alive. Other family members and his grandchildren did not find out the truth until it was revealed by a historian in 2005.[40] Kehoe was in such demand as a consultant to corporations that starting in 1929, Ethyl, GM, DuPont, Frigidaire, and others gave $100,000 (worth several million today) to his Kettering industrial toxicology laboratory at Cincinnati to do research. But they required that all the work be submitted to, and vetted by, the corporations. Even in 1965, after his retirement, Kehoe warned staff at the Kettering lab that the studies and papers must be kept secret. As a later head of Kettering put it, Kehoe ran the lab like "the Medical Department of the Ethyl Corporation."[41]

By the end of World War II, Kehoe had become a captain assigned to the OSS and sent to Germany with other intelligence agents just two months after V-E Day. As Davis notes, he interviewed key German scientists and brought back critical studies on topics ranging from chemical warfare to pesticides, pharmaceuticals, and industrial materials.[42] Among them were studies done by I. G. Farben, the notorious industrial conglomerate that used slave labor. Farben had been aided in producing leaded gasoline that powered the Nazi war machine based on secrets shared with them illegally before the war by Standard Oil and Ethyl. Harry Truman, in investigating American corporate cooperation with Hitler in the late 1930s, called it treason. These Farben studies and others that Kehoe obtained revealed that the production of dyes did indeed produce bladder cancer in dye workers, just as Wilhelm Hueper had tried to warn. But all this work was kept secret and unknown to the public or the medical establishment. Where did it go? Devra Davis asks and answers the question. "Detailed summaries of Nazi research on the hazards of tobacco and various chemicals made their way to executives of many of the U.S. corporations then producing these materials."[43] And, incredibly enough, the paths of Hueper and Kehoe crossed a number of different ways. Hueper had asked for an autopsy of a baby he suspected was killed by lead poisoning from the Chamber Works in 1936. The report was sent to Kehoe instead. But more important, as the result of a court case, Hueper discovered in 1960 that Kehoe, working in behalf of corporations, had been reviewing his work for years and continuing studies that Hueper had been ordered to discontinue. Benzidine dyes had become a mainstay of DuPont production, and Kehoe kept close track of their effects on workers. No cancer registries were available at the time. Only corporations had such information. The DuPont Corporation disclosed no problems with bladder cancer at their facilities for decades. Health problems and cancer deaths, watched by Kehoe, were proprietary secrets. Only in 1980 did it become known that since the factory opened and Hueper issued warnings, 364 cases of bladder cancer had occurred.[44]

The Search for Cures but Not Prevention

Meanwhile, American faith in technology and the success of World War II had led to the belief that technology could conquer everything, even cancer. The chemicals produced in World War II, as well as radiation, when controlled and aimed at human tissues, could reduce and sometimes kill cancer tumors. And there was profit to be had in producing, deploying, and analyzing the results of such wonder-working techniques. Thus when Richard Nixon, once again, was looking to burnish his medical credentials against a possible challenge from Senator Edward Kennedy, he signed up and went all out for the famous "War on Cancer." Research boomed and money flowed. The American Cancer Society agreed we must have a cure for cancer, no matter what the cost. A minority of doctors, researchers, and others disagreed, says Davis, but they were soon ignored. Years later a few people like Joshua Lederberg of the Rockefeller Foundation were able to send her curling old op-eds decrying the excessive focus on and funding for cures and not prevention. But mammography, radiation treatments, chemotherapy, all were pushed out front. They did indeed stave off the worst or saves lives, but the waves of cancer incidence still grew larger.

Despite the revelations of *Silent Spring*, it was on this unequal playing field—with the environmental causes of cancer ridiculed and suppressed and noted experts vouching for chemicals produced by corporations who paid for their research—that contemporary environmentalists entered the fray. Devra Davis shows in painful detail how advances that could have saved thousands of lives were held back by leaded gas, tobacco, the failure to endorse Pap smears, and more. Only with the slow growth of a stronger environmental movement and its medical allies does Davis begin to see some hope. She begins her final section with an epigraph from Harriet Hardy, Alice Hamilton's friend and Harvard physician colleague, who inspired a new generation of physician activists like herself. "All scientific work is liable to be upset or modified by advancing knowledge. That does not confer upon us a freedom to ignore the knowledge we already have, or to postpone action that it appears to demand at any time."[45]

Davis concludes with some ongoing, contemporary battles over the dangers of other materials, including vinyl chloride. Once again, vinyl chloride was first produced by the Germans, using chlorine, the same poisonous gas they used on the battlefields of World War I. Experience with vinyl chloride led to Saran Wrap, as well as more important products made from polyvinyl chloride (PVC), such as plastic pipes, which are stronger and more flexible than those made of steel or lead. They could only be dissolved by benzene. Otherwise, PVC is almost indestructible. It was assumed that vinyl chloride, unlike chlorine itself, was not harmful. It had many wonderful uses, such as a propellant in cans of hair spray. But as, Devra Davis explains, this assumption, too, proved to be wrong. Enter Judy Braiman, one of a number of women that Davis admires, who refused to accept the bland assurances of corporate and medical experts. She just set out to organize. Davis describes her as a "small, strong woman, with the tough edge of a grandmother who knows how to throw a football." She is "an improbable revolutionary."[46]

Judy Braiman and the Dangers of Vinyl Chloride

In 1966, as a thirty-year-old mother in upstate New York, Judy Braiman had three small children, was married to a successful lawyer, and was beginning to have thinning hair. Told by her hairdresser about a new hairspray that increased volume, she sprayed her then-fashionable bouffant hairdo several times a day in a small bathroom. She soon developed a severe cough, her ribs ached, and she could barely walk up stairs. She was losing weight and spitting blood. Her doctor, William Craver, told Davis forty years later that he would never forget her X-rays, filled with lesions: "They showed patchy, fluffy infiltrates scattered throughout both lungs." It looked like choriocarcinoma, a rare and fast-growing cancer. Braiman went into an operation prepared to die. After a lung, a rib and some muscle were removed, it was discovered it was not cancer after all. Her lungs had sixty different deposits of small round hairspray modules that had set off lesions called granulomas. There were also fatty globules found by Dr. Craver. It turns out that only one hairspray manufacturer, an outfit named Bonat, used fat in its product. Braiman became an activist. She talked to the press, including the well-known reporters Jack Anderson and Les Whitten, who produced a widely read column on her case. She organized in her community, spoke to congressmen, wrote letters, did all she could to reduce the toxic materials in hair care products. Her efforts were getting noticed when she received a letter dated December 17, 1973, from the Bonat Company threatening to sue her for slander if she mentioned on her next television appearance that her personal injury was caused by their product. By this time, cans of hairspray were about one-half vinyl chloride gas. But reports were now circulating about the danger, and industry quietly began to discourage the use of vinyl chloride. Clairol Corporation told a reporter that its popular Summer

Blonde Hairspray would no longer be made with vinyl chloride. The AP reporter, John Stowell, was outraged when he leaned that before making its claim, Clairol had produced, in less than a month, an entire year's supply of Summer Blonde still containing vinyl chloride.[47]

As it turns out, the dangers of vinyl chloride had already been observed and reported even long before Judy Braiman took action. A bone condition of stunted fingers, or acro-osteolysis, had been observed in individual production workers in the 1940s in France; it was rare and not yet linked to vinyl chloride. But by the 1960s in France and Italy acro-osteolysis began to be reported in groups of vinyl chloride workers. In 1964, in the United States, an industry physician, Dr. Rex Wilson, began to observe stunted fingers in workers and wrote about it to his counterpart at another vinyl chloride plant in Ohio. The letter was marked "confidential." It describes the affliction carefully and asks that a Dr. J. Newman make quiet, careful observations of the hands of workers at his plant. He makes it clear that this is a top priority, but must not be discussed. "Will you please advise me by January 1, the approximate number of hands that you have seen and of any positive findings." By 1967, the results of Dr. Newman's quiet investigation were clear. Out of three thousand workers at the Avon Lake factory, thirty one cases of this previously rare disease, fully one-third of all those known in the world at the time, had been discovered. This led to animal studies.

In 1971, the Italian scientist Paulo Viola found cancers of the skin, lungs, and bones of rats exposed to vinyl chloride. Montedison, a major Italian manufacturer of vinyl chloride, was aware of these studies and commissioned Cesare Maltoni to begin a series of studies. In his tests, Maltoni exposed rats to very small amounts of vinyl chloride, but for far longer than usual. It approximated the sort of exposure a woman like Judy Braiman might get from hairspray or a factory worker might get on the job. The results showed one in ten of the exposed rats got a rare tumor of the liver known as angiosarcoma—an always fatal cancer. But all these studies were, of course, unknown to any government. By 1972, Maltoni was convinced that vinyl chloride was a serious human health threat. He followed the rules of his contract, however, and kept quiet, assured that his results would be made known in due time. Maltoni assumed this meant that the company would reveal these hazards in meetings scheduled with the Italian, French, and American governments. When his findings were not presented, he was furious and, in 1974, published the results in the medical journal *La Medicina del Lavoro.* More evidence emerged, including reports in the United States. Finally, in 1975, OSHA severely restricted exposures to vinyl chloride.[48]

But even decades later, when Gerald Markowitz and David Rosner, the authors of *Deceit and Denial* (2003) wrote a chapter on the sordid history of vinyl chloride and revealed evidence of an illegal conspiracy to conceal evidence among corporations like Dow, Monsanto, B. F. Goodrich, and Union Carbide, they were slapped with a lawsuit. So was the publisher, the University of California Press, and subpoenas to turn over files were sent to the academic reviewers and to the Milbank Memorial Fund that had supported the work.[49] Even one of the world's top epidemiologists, Sir Richard Doll of Oxford, challenged the findings of a 1987 assessment report from the International Agency for Research on Cancer. It had updated a prior 1979 report and concluded that, in addition to angiosarcoma, vinyl chloride caused other kinds of liver cancer, plus cancers of the brain, lung, and bone marrow. Doll did not dispute that vinyl chloride caused angiosarcoma, but argued that it was not a cause of more common cancers of the brain and liver. Doll's review was limited and flawed, nor did it mention that he had done the study as a paid consultant for the Chemical Manufacturers Association. Nevertheless, vinyl chloride workers who developed the more common tumors of the brain, liver, and lung could no longer win compensation in court. After his death in 2005, a letter was found addressed to Doll from Monsanto, revealing that he had been a consultant since at least 1979 with a fee of $1,500 per day.[50]

Devra Davis, Alice Hamilton, and Benzene

It is appropriate that Devra Davis's final example of the delay and destruction caused by corporate obfuscation of science, secret research, and opposition to regulation involves her admiration for Dr. Alice Hamilton. In describing the ongoing attempts to control and regulate benzene, still an issue in our time, Davis begins with Dr. Hamilton, reviewing in some detail her articles on the adverse health effects of benzene from 1916 and 1917 that reference work in German, French, and other journals, even cases reported at Johns Hopkins in 1910. Davis describes the young women whom we met in chapter 2 at a bicycle tire factory in Sweden who hemorrhaged and died in 1897. She recounts Alice Hamilton's visits for the U.S. Bureau of

Standards to forty-one plants making explosives with benzene-based compounds. Again Hamilton reported poisoning and death from massive hemorrhages. Between 1919 and 1940, at least thirty-three publications, many by Alice Hamilton and the National Safety Council, advised replacing benzene with safer solvents. Yet by 1948, after the war, the American Standards Association, composed of experts from industry, held that workers could be exposed safely to 100 parts per million (ppm) of benzene over an eight-hour period. Even after *Silent Spring*, Earth Day, and the political conversion of Richard Nixon to environmentalism and the EPA, corporate and conservative obstinacy continued. In 1970, the American Petroleum Institute asked Bernard Goldstein of NYU's Institute of Environmental Medicine to report on the current world literature on benzene. When Goldstein came back with the obvious answer that benzene causes leukemia, he told Devra Davis, "API refused to fund us." Even more telling is the experience of Marvin Legator, a leading advocate for Americans exposed to toxic chemicals, with whom I worked in the 1990s.

Legator had started out with Shell Oil in the 1950s and then joined the federal Food and Drug Administration. An honest, eager, and enthusiastic researcher, Legator founded in 1969 an early group of environmental health specialists interested in the effects of chemicals on genes. Called the Environmental Mutagen Society (EMS), it included Alexander Hollander, Joshua Lederberg, and Samuel Epstein, all then at Harvard. At a meeting of EMS at Brown University, Legator was introduced to Jack Killian, the medical director of Dow Chemical, who said he wanted to carry out research by monitoring workers on a regular basis. Legator was impressed and saw an opportunity to see if benzene was getting to the bones of workers. Dow Chemical, of course, was under fire at the time as the producers of napalm, the sticky, jellied gasoline incendiary being widely used in Vietnam. Epstein, to this day a fiery leader of coalitions to link the environment and cancer, was furious. Nonetheless, Legator jumped at the chance and took a well-funded consultantship with Dow Chemical to do toxicology at the University of Texas Medical Branch. He became expert and well known as a leader in looking at the DNA of workers. But he, too, went too far. When his research finally revealed that benzene was causing damage to chromosomes, Dow Chemical pulled the funding plug. In the twenty-first century, after Legator had died in 2004 of cancer brought on by chemical exposures, the case of benzene is still not fully settled. Davis describes how, despite

OSHA standards, there has been little enforcement in recent decades since the days of Jimmy Carter; OSHA works in close collaboration, not with independent researchers, but with the industry group, the American Chemical Council.

And even as far more thorough and widespread studies that document wider dangers from benzene have been carried out in China, in cooperation with the NCI and Berkeley, the results are being challenged by a $27 million campaign from Exxon Mobil, British Petroleum, Chevron-Texaco, ConocoPhillips, and Shell.[51] Benzene has been seen as a hazard to human health from the days of Alice Hamilton and World War I. It was recognized as a cause of leukemia as Rachel Carson was finishing her graduate work and teaching at Johns Hopkins in the 1930s. It was severely restricted by OSHA and banned in places overseas throughout the time that Devra Davis and I and others were at Hopkins and advocating in the 1980s, 1990s, and beyond. It is still dangerous today. But, it seems, some still believe more research is needed.

Davis's calling is in reaching out beyond a small cohort of scientific specialists and even well-informed environmentalists. At every turn, she makes the history of advances in environmental health and obscure studies relevant and understandable; she tells us what they mean. She remains on the forefront of environmental health issues today with impassioned advocacy and writing on the cancer dangers inherent in the widespread use of cellular telephones.[52] And, like the women whose path she tries to follow— Ellen Richards, Alice Hamilton, Harriet Hardy, Anna Baetjer, Mary Amdur, and Rachel Carson—Devra Davis knows that epidemiology and environmental health are not really about statistics. They are about saving human lives, about individuals and their stories, about people, women and men, like you and me.

7 ⬧ RACHEL CARSON AND THEO COLBORN

Endocrine Disruption and Ethics

I AM HOOKED FROM THE START by the picture of the double-crested cormorant with a grossly deformed, twisted bill. More than piled up data, more than scatter plots in which only scientists see patterns, the effects of toxic chemicals on such a superb flying, swimming, and diving specimen are inescapable. It helps that the pictures are being shown by Theodora Emily Colborn, always known, despite her doctorate, as Theo. She has breezed into the board room at Physicians for Social Responsibility in Washington, bringing with her a fresh gust from the West. Other pictures appear. Hermaphroditic beluga whales, the cute white ones that kids love to watch swimming in aquariums. A Canadian wildlife service ranger had told me a decade before while my wife and I were birding and whale watching at the confluence of the Saguenay River and the mighty St. Lawrence in Quebec that they were filled with PCBs and other chemicals. If buried, they would qualify under Canadian law as toxic waste.

Colborn is upbeat, friendly, almost chipper, despite the gloomy subject. There are none of the ways of Washington about her. No self-aggrandizement, no stentorian, pompous, definitive tones. She seems dressed for the outdoors—plain blouse, simple pants, a carefree, bowl haircut that looks as right on her as it did on my daughters when they were kids.

What is amazing, as she talks, is the unusual combination of sweeping science and simplicity. She reels off a string of animal studies—Soto and Sonnenschein, vom Saal, the Jacobsons, and more. This is becoming a very pleasant morning away from the clamor of Capitol Hill. I learn more than I have in years. I like Theo Colborn immediately. We are fellow

FIGURE 7.1   Theo Colborn, who has revealed
endocrine disruption in wildlife and humans.
Michael Collopy Photography.

birders, talkers, generalists. There is nothing angry or scolding in her talk.
This woman, unpretentious, unthreatening, I think to myself, should talk
to oodles of Americans. She is one of them. Maybe they would listen. As it
turns out, I am right that many, many Americans will react to Theo Colborn
as I just have in 1992. But I am wrong, quite wrong, that she will not seem
threatening. I am watching a former pharmacist and western sheep farmer
fire an opening salvo in a new phase of the environmental revolution.

Like Rachel Carson before her, Theo Colborn is best at integrating and
communicating large amounts of new, cutting-edge science, in bringing
a fresh, keen eye and imagination to the meaning of far-flung, often little
noticed, forgotten facts. And beneath the easy-going, self-effacing surface,
there is determination, even doggedness, that will not be deterred, that
keeps her working deep into the night. Colborn's career is unconventional,
starting late in life. When she first briefed the staff at PSR, she was already

in her sixties, a widow who had completed her PhD in biology at the University of Wisconsin just a few short years before; a woman who raised four children and helped her husband with a chain of drugstores and sheep in Colorado before going back to school. Theo Colborn, like many women we have seen, does not get the automatic deference that comes with early success and supersized credentials, high positions in academia, government, even NGOs. There are no classmates or cronies to welcome her inside the doors of influence and power. She is not of the establishment; it is both a blessing and a curse. She is not bound by long-held theories and positions. She can and does ask new, refreshing questions. She can leap over disciplinary boundaries and departments, ignore the conventional wisdom when it seems wrong, or does not fit, develop new concepts, new rules, new ethics.

Yet when her work challenges scientific and social orthodoxy, the backlash is intense. But like Carson and others, Theo Colborn does not really face abuse alone. She is part of a wide network of friends and colleagues, part of the broader environmental movement that now is truly global. Colborn is part of our emerging portrait that includes the work on water quality and health and consumer safety of Ellen Swallow Richards, the love of birds and wildlife stirred by Florence Merriam Bailey, and the ecological ethic and broad human rights concerns of Rachel Carson. Like her younger contemporaries, Terry Tempest Williams, Sandra Steingraber, and Devra Davis—who have built on the work and values of Rachel Carson to expand ecology to embrace nuclear power and weapons, the genocide of animals and Africans, the dangers of chemicals in communities, and the failure of both corporations and our government to protect us from environmental cancers—Theo Colborn explores and expands the full implications of Rachel Carson's work. After *Silent Spring*, connecting the environment to the causes of cancer became a primary concern. Rachel Carson, of course, had broader interests, too. But in her day, the science of chemicals that change hormones was not yet fully developed. Carson focused mostly on the genetic and cellular damage leading to cancer that pollutants can produce. She alluded to hormonal changes and speculated that there may be wider dangers here. But it would take Colborn to add new chapters on endocrine disruption to Rachel Carson's work.

Beneath the picture of that horribly hobbled cormorant is a revolutionary idea. Exposure to even tiny, barely traceable amounts of persistent pollutants in the bodies of birds and other animals, even human beings, can mimic, replace, or upset the role of vital hormones that control reproduction, gender formation, growth, learning, and certain kinds of natural behavior. And it is not always the size of the exposure, but the timing that matters. After Theo Colborn, this new, widespread environmental health threat will be known as endocrine disruption. If the implications of this science are fully accepted by society, it will mean changing almost every product that we now manufacture and our social values, too. Along with the global threat of climate change, brought on by burning fossil fuels, the adverse effects of many common synthetic chemicals on wildlife and on people is one of the most critical threats to humanity. How endocrine disruption came to be discovered and presented to the public by Theo Colborn and her colleagues is like a detective story or a Western movie where the new, innocent, upright marshal has just ridden into town.

Theo Colborn—Unlikely Hero

Theodora Emily Decker Colborn seems, at first, an unlikely hero. She was born in Plainfield, New Jersey, now a prosperous New York suburb, on March 28, 1927, when Plainfield was still rural. Her parents, Theodore Decker and Margaret LaForge Decker, had a farm in the hills above Plainfield where young Theo was fascinated by and played in the river that ran nearby. She developed a love of the outdoors and nature at a time when writers like Florence Merriam Bailey, Olive Thorne Miller, and Neltje Blanchan were all the rage. Theo would beg to always be outdoors and to feed the birds. In fifth grade, she had a teacher who would take her and the other students on 5:00 A.M. trips to Branch Brook Park to hear the first morning bird songs and to study nature. By high school in East Orange, Theo began to blossom academically when she discovered chemistry and studied botany and other sciences. But still a tomboy, she sought the outdoors whenever possible; she frequently went birding with her close friend, Marion Wohlhieter, whose family had a summer place on Lake Lenape. Theo was able to go on to Rutgers University thanks to a scholarship, where she earned a degree in pharmacy in 1947. On campus, she met Harry R. Colborn; the college sweethearts married on January 20, 1949. Harry took over his father's drugstore in Newton, New Jersey, and expanded the business to three busy stores. Theo gave birth to four children, Harry, Christine, Susan, and Mark.

But throughout her time in Newton, she continued to bird, became active with the Audubon Society, helped create an early bluebird box trail, and joined the fight against the Tocks Island Dam that was designed to bring water to New York City, but whose reservoir, if created, would flood both homes and environmentally important areas.[1]

By 1963, just before Rachel Carson died, the couple sought a simpler lifestyle and, like many, moved out West. The Colborns raised sheep in western Colorado and owned pharmacies as well. In leaving New Jersey for fresh air, clean water, and a better lifestyle in Colorado, Theo Colborn, perhaps unwittingly, journeyed directly into the mountain heart of a proud western women's environmental tradition.

Colborn and Western Environmental Women

Colborn was following now in the footsteps of Martha Maxwell, who had found and collected new species, opened a trail-blazing natural history museum in Boulder, and been a star, with her gun and naturally mounted birds and animals, at the 1876 Centennial in Philadelphia. The mountain West also drew other rugged, unconventional women to its wild and open spaces. Even before the Civil War, a pioneering feminist, the twenty-year-old Julia Anna Archibald Holmes, had traveled in 1858 by ox cart with her husband and brother eleven hundred miles from Kansas to Colorado where she became the first woman ever to climb Pike's Peak, wearing what she called "the American costume"—bloomers. In the 1880s, Virginia Donaghe McClurg explored the Mesa Verde cliff dwellings, as had anthropologists Alice Fletcher and Matilda Coxe before her. McClurg then lectured widely about Mesa Verde throughout the 1880s. In 1900, after teaming with the Colorado Federation of Women's Clubs, where she headed a committee to "save the Cliff Dwellings," McClurg led a six-year-long campaign aimed at Congress that led in 1906 to the creation of Mesa Verde National Park.[2] In 1916, Esther Burnell spent the winter as a homesteader in Estes Park. There she made a thirty-mile solo trip by snowshoe across the Continental Divide, gaining a reputation as an outdoorswoman. The next year, despite the agency's record of not hiring women, Esther and her sister Elizabeth took the U.S. Park Service exam and became licensed forest rangers. Elizabeth ran a trail school and led many groups up the challenging heights of Long's Peak.[3]

Women mountain climbers in Colorado were part of a broader environmental movement, too. The Colorado Mountain Club (CMC), formed in 1912, promised in its charter "to stimulate public interests in our mountain area" and "to encourage the preservation of forests, flowers and natural scenery." One of the club's cofounders was Mary Cronin, a teacher, who went on to climb all of the forty-eight 14,000-foot mountains then known in the Colorado Rockies. Laura Makepeace of the Fort Collins CMC noted that the group was committed to the preservation of nature: CMC members saved trees, built nature trails, created bird sanctuaries, supported the founding of Rocky Mountain National Park, and advocated for Dinosaur National Monument, the removal of billboards in Denver Mountain Parks, and led a campaign to "Save the Wildflowers."[4]

Since the 1870s, Colorado had also become a haven for those suffering from eastern, urban afflictions. Its fresh air and clean water were touted as cures for illnesses like tuberculosis. Boulder opened its Colorado Sanitarium in 1896 and welcomed people "suffering from any chronic disease." As the twentieth century opened, the Colorado Promotion and Publicity Committee listed the sanitarium as a main attraction, along with numerous health resorts. "If you wish to breathe air that is death to the bacilli of tuberculosis, come to Colorado!" was the pitch.[5] It was just such a desire to restore her health with western air that had taken Florence Merriam Bailey to the West so many times. The results were so successful that she could travel and hike throughout the West and produce her classic, groundbreaking *A Handbook of the Birds of the Western United States* and other works.

Water, too, was an issue that had consumed western environmental women like Mary Austin, who wrote of the glories of native, arid deserts. Austin fought needless western irrigation as a representative on the Colorado River Commission in 1927, opposing and writing about the damming of the Colorado River to meet the growing, seemingly insatiable thirst of metropolitan Los Angeles.[6] By the early 1950s, when Theo Colborn was having children and working in the pharmacy, Rachel Carson was already deeply engaged with members of the Wilderness Society in a battle to prevent a dam that would flood the national monument areas of Colorado and Utah, including the Dinosaur National Monument that had been fought for by women only thirty years before. David Brower, as the new and first national director of the Sierra Club, is often highlighted for his emerging leadership in this lobbying effort. But lesser-known women were also actively engaged in directing these efforts. The first emergency meeting to plan strategy to save the Dinosaur Monument was held at the Berkeley home of Doris Leonard and her husband, Richard. Several women also made the initial trip to the monument on the Colorado border to gather firsthand accounts and remained active in the campaign throughout. They included Polly Dyer of the Mountaineers; Cicely Christy, a landscape architect; mountain climber Marjorie Bridge Farquhar; and Kathleen Goddard Jones, for whom organizing around the Dinosaur Monument led to a career of national activism with the Sierra Club, the Audubon Society, and the Native Plant Society.[7]

Theo Colborn's Early Activism and Studies of Great Lakes Pollution

But by the 1970s, oil shortages had led to coal mining on a massive scale, along with other minerals, in the valley near the Colborns' sheep farm. The mining was rapidly polluting the local river, Gunnison's North Fork. Before long, Colborn, who had left the urban East, became deeply involved in western water issues. At the same time, she maintained meticulous records on a weekly and annual basis on the birds of Colorado. She drew the attention of the Audubon Society and became quite well known. For the League of Women Voters, she debated Wayne Aspinall, the conservative member of Congress responsible for a number of dams and the arch-nemesis of national groups like the Sierra Club. As a result of such environmental work, Colborn says, "I got a call from Governor Dick Lamm to join his Natural Areas Program, which he was just starting up. I served on that panel for about twelve years."[8] During that time, while also fighting the Four Corners coal-fired utility in the southwest corner of Colorado, Colborn made the decision to go back to school for an master's degree at Western State College. While there, she also studied at the affiliated Rocky Mountain Biological Laboratory. She became friends with and learned much from a number of renowned scientists at this world-class facility, including Paul Ehrlich of Stanford and John Holdren of Berkeley. She also found congenial female scientist friends and colleagues, such as biologist Svata Louda, known for her studies of insects in water, and Sherry Holdren, then completing her PhD. For her master's thesis, Colborn researched the effects of metals like cadmium and molybdenum, introduced into the water because of mining, on small invertebrates.

Then, about a week before her thesis defense, she got an early taste of the tactics of polluting corporations. Colborn observed the head of operations for the AMAX Corporation, which wanted to mine nearby Red Lady Mountain, enter her building and head to the office of one of her thesis committee members. He was clearly holding in his hand one of the five copies of her thesis that she had personally bound in blue. She informed her adviser and the department head about this breach of academic integrity. At her defense, as the committee members were about to query her, the professor who had leaked her thesis to the mining corporation said, "Before we begin, I want to ask what the effect of this thesis will have on my associates from AMAX up on the mountain." Colborn's adviser, Marty Apley, immediately halted the defense and went into executive session with the committee. When they emerged, Apley told her that they were approving her thesis immediately and recommending her for further PhD work.[9]

With her four children grown and her husband, Harry, having died, Colborn got a fellowship and assistantship to teach and complete her doctorate in zoology at the University of Wisconsin, with an emphasis on epidemiology, toxicology, and water chemistry. It was at Wisconsin that she first read Rachel Carson and Aldo Leopold and became familiar with the work of Joe Hickey on pesticides that Rachel Carson had cited. As she was finishing, one of her students, who had been at the Office of Technology Assessment (OTA) in Washington, DC, told her she should apply for a fellowship there. Colborn was eager to do it because she had carefully reviewed many, many OTA documents before debating Wayne Aspinall. And so, at age fifty-eight, she received a fellowship from the Office of Technology Assessment and, for the first time in her life, set foot in the nation's capital. Two years later, as she was about to head back to Colorado, she got a call asking if she would begin work for the Conservation Foundation to provide the science for a new research project and book that had been requested by the U.S. and Canadian International Joint Commission (IJC). The IJC oversees environmental conditions in the five Great Lakes, the largest repository of freshwater in the world.[10]

Theo Colborn's work consists of a large body of scientific, peer-reviewed articles and her breakout 1990 work, *Great Lakes, Great Legacy?* As her work was finishing with the Conservation Foundation, part of the World Wildlife Fund, as Colborn tells it, "Providence was again at work." She went to a meeting of the Audubon Society of Fairfax, Virginia, in suburban Washington to hear Pete Myers, vice president of the National Audubon Society.

He was lecturing on pectoral sandpipers and their mysterious decline. In a crowd that contained many luminaries like Bruce Babbitt, secretary of the interior, Colborn was unable to speak to Myers. But she called him up later and explained that she believed the sandpiper's decline was related to chemical exposures and endocrine disruption. Myers understood the implications immediately and asked her to come right away to New York to brief Audubon president Peter Berle and his staff. Berle quickly asked her to join the Audubon staff, but Pete Myers then found out that he had just been picked to create and head an $80 million project on the environment for the W. Alton Jones Foundation in Virginia. He wanted Colborn to join him, with a paid chair. Colborn agreed and was able to stay on in Washington at the World Wildlife Fund (WWF) with full pay from W. Alton Jones.[11]

While at WWF, Colborn was approached by an official of the S. C. Johnson Foundation of Racine, Wisconsin, who asked what she would most like to accomplish. Colborn answered that she wanted to create collaboration between researchers in both wildlife and human health. Unaware of each other's work, they had been working separately on the mystery of some connection between chemical exposures and disruption of the endocrine system. From this conversation grew a meeting at the Johnson Foundation's conference center, Wingspread, in Wisconsin, where important groups of experts have gathered over the years to issue innovative intellectual and policy pronouncements. Colborn then assembled and organized the top researchers; they quickly realized that the changes they had been observing independently in human and wildlife studies were closely related. Colborn and her colleagues then issued the Wingspread Consensus Statement on the dangers of endocrine disruption; it was followed in 1992 by *Chemically-Induced Alterations in Sexual and Functional Development: The Wildlife/Human Connection*, a pioneering collection of scientific and technical articles based on the conference, which Colborn pulled together and edited.[12]

Our Stolen Future—A Detective Story That Makes Endocrine Disruption Understandable

Soon after, Colborn was offered a $50,000 grant by Shelley Hearne of the Pew Charitable Trusts, who had a PhD in environmental health from Johns Hopkins, to produce a popular version of her pioneering work. She labored through eight chapters, but publishers found the manuscript still too technical for a general audience. So Colborn turned to Pete Myers, who knew the science, and, as she puts it, "is a wonderful writer." But months passed. Myers was simply too busy heading up projects and running the W. Alton Jones Foundation to make much progress. Then, at an annual conference of the Society of Environmental Journalists that she and Myers were attending, Colborn met Dianne Dumanoski, an environmental reporter with the *Boston Globe*. At dinner, Colborn persuaded her to join their project and rework the material. Dumanoski, who had followed Colborn's work very closely over the years but had never been permitted by her editors to write about it, was in tears at the opportunity to write freely, based on her own convictions.

But then, as she neared the conclusion of the book, Dianne Dumanoski was diagnosed with breast cancer. Nevertheless, like Rachel Carson with *Silent Spring*, even while undergoing treatment, she would not be deterred from finishing what had become her life's mission. Between the team of Theo Colborn, Pete Myers, and Dianne Dumanoski, the popular book *Our Stolen Future* was born. Its appearance and reactions in the media caused a sensation; the danger of endocrine-disrupting chemicals was about to catapult into the mainstream. Theo Colborn had suggested two devices for the book that made its impact even more powerful and memorable than the most skillful telling of the facts. As with Rachel Carson's "A Fable for Tomorrow," the fictionalized, but fact-based picture of a town devastated by pesticides that opens *Silent Spring, Our Stolen Future* features a creative nonfiction narrative designed as a detective story. In it, each actual scientific clue and wildlife calamity leads the protagonist, Theo Colborn, presented in the third person, on to yet another mystery to be solved. And, in perhaps its most memorable feature, *Our Stolen Future* has at its center a dramatic, almost incredible composite story, based on actual scientific reports. We follow a single molecule of an endocrine-disrupting chemical, one of a number of kinds of polychlorinated biphenyls, or PCBs, on a long, complex journey from a broken transformer in Texas to the breast milk of an unsuspecting Inuit mother in the far-off Arctic.[13]

The detective story in *Our Stolen Future* opens in 1952, ten years before *Silent Spring*. Rachel Carson is still known as the best-selling nature writer of the *Sea around Us*, whose prose captures the mystery and depths of the ocean. Charles Broley, a Canadian banker and amateur ornithologist, monitors eagles in the Everglades. It is he who will report the first in a series of unexplained wildlife disasters in the opening chapter, labeled "Omens." He has been watching eagles carefully since 1939 when the staff of the National Audubon Society first suggested he keep track. Eagles were nesting all along the west coast of Florida, he reported in the early forties, from Tampa to Fort Myers, with some 125 active nests in all. Broley also clambered up and banded about 150 new eaglets every year. Then, after the war, starting about 1947, everything changed. Eagle numbers plummeted, and Broley saw increasingly bizarre behavior among his eagle pairs. They still gathered sticks in early winter and added them to their growing, sprawling nests. But two-thirds of the birds showed no interest in any other parts of established courtship and mating rituals. The eagles, in Broley's words, just sat around and "loafed." What could be the reason? At first, Broley attributed their lack of mating interest and activity to the disturbance of the encroaching housing developments of Florida's postwar housing boom. University researchers concurred. But Broley began to question this explanation as he kept up his faithful observations. By the mid-1950s, he had become convinced that fully 80 percent of Florida's eagles were, for as yet some unexplained reason, now sterile.[14]

These mysteries pile up like dead alewives on a lake shore at the start of *Our Stolen Future*. With postwar affluence, the demand for mink coats surged, but the mink farms along Lake Michigan collapsed by the early 1960s. Mink, who live off fish and were fed them on commercial farms, continued to mate, but stopped having pups. Only those fed with more expensive fish imported from the West Coast survived. Researchers from Michigan State University finally tie the problem to the pollutant PCB in Great Lakes water. There had been a similar mink population crash a decade earlier from the synthetic hormone DES. But what could possibly be the cause and the connection?[15]

Other seemingly unconnected mysteries and maladies occur. Biologist Mike Gilbertson observes unhatched eggs and abandoned nests all along the breeding grounds of Herring Gulls on Near Island in the Great Lakes. Worse, there are dead chicks everywhere, with grotesque deformities—club feet, missing eyes, twisted bills. It appears to be from chick edema disease, which Gilbertson had studied while in England. It could be from exposure to dioxin, he surmises. As Colborn puts it, "Gilbertson's colleagues and superiors greeted this theory with skepticism bordering on derision. Dioxin had never been reported in the lake and the gull eggs showed no dioxin that they could measure. Nevertheless, Gilbertson remained convinced that dioxin was somehow the contamination killing off the chicks."[16]

Elsewhere, reports began to filter in, from the late 1960s onward, of excess eggs in the nests of Western Gulls. Then, over the next two decades, come increasing reports of same-sex pairings with excess eggs and no chicks from Herring Gulls in the Great Lakes, Glaucous Gulls in Puget Sound, and among endangered Roseate Terns off the coast of Massachusetts. There were massive die-offs from viruses among seals in the North Sea. But the eighteen thousand victims that were studied showed different virus symptoms in different locations—a sign that something in the water was affecting the immune system of these seals. And in Florida's Lake Apopka, a chemical spill in 1980 had killed off 90 percent of the alligator population. Nearly a decade later, when the lake no longer showed measurable traces of chemicals, the alligators were back. But only 18 percent of the alligator eggs laid were hatching. Half of the alligators that did hatch died within ten days. And when Louis Gillette of the University of Florida investigated, he also found that of the male alligators alive in the lake, more than 60 percent had tiny, micropenises.[17] What was going on? Striped seals in the Mediterranean were dying off from virus distemper. Once again, researchers found elevated levels of PCBs in the dead seals and suspected that this industrial lubricant might somehow be the cause. And among human males, Niels Skakkebaek of Denmark was finding triple the rate of testicular cancer and plunging sperm counts. He began to review records worldwide and got similar results. Average human male sperm counts had dropped 50 percent between 1938 and 1990. Because of the relatively short time frame, genetic factors were ruled out. Some sort of environmental cause seemed to be the answer. More and more reports of unusual, even bizarre, problems like these began to surface around the globe. "Each incident was a clear sign that something was seriously wrong, but for years no one recognized that these disparate phenomena were all connected. While most incidents seemed linked somehow to chemical pollution, no one saw a common thread. Then in the late '80s, one scientist began to put the pieces together."[18]

There were thousands of such pieces. When Theo Colborn began to assemble them for the Great Lakes study with her supervisor, Rich Liroff, the reports just kept piling up. Governors had declared the Great Lakes "cleaned up." The limits on what waste could be poured in the water and the virtual ban on DDT in 1972, following *Silent Spring*, had indeed made a difference. But Colborn had a gut feeling from all the animal studies that things were still not right. Faced with forty-three boxes of scientific studies, with more being flicked by the post office underneath her door each day, Colborn decided to focus on cancer. There were cancer registries, recently set up, in the region. There were cases of cancer in fish, and "[t]here was a strong consensus among many people living in the region that they were being exposed to higher levels of toxic chemicals than those living elsewhere and were suffering from higher than normal cancer rates." But even the focus on cancer proved difficult, because the fish in the Great Lakes were exposed to so many different chemicals. And alternative explanations still being offered had to be examined—viruses or other nonchemical causes.[19]

Finally, Colborn found connections at an early conference on chemicals and health in Toronto, Canada. Researchers had now determined that fish showed more cancer when feeding below discharge pipes; bottom feeders, like catfish, that spent their time amid the mud and sediments with discharges fared the worst. Studies had also now linked contaminated sediments to cancer by using confined fish in the lab. The link was also made to PAHs, or polyaromatic hydrocarbons, produced by the incomplete burning of any carbon-containing substance, from "gasoline to hamburgers on the outdoor grill." The biological mechanism for all this was also becoming clearer, Colborn found. "Diseased fish showed changes in the liver caused by bonding between carbon-based organic chemicals such as PAHs and the DNA in the nuclei of their cells." There were also reports at the Toronto conference of reduced testes size in fish exposed to organochlorines, but which ones remained a puzzle. Colborn kept this intriguing, troubling fact in the back of her mind, but kept after the contamination–cancer link.[20]

But a wall was hit when the cancer registries proved to be too new to be of any use. Colborn then turned to broader cancer reports, "poring over the computer printouts and reports for hours, analyzing the data from various perspectives to see if she could tease out meaningful patterns." Nothing emerged. She tried looking for clusters of a single cancer, geological patterns, anything. After months, she came to the conclusion "that no matter which way she cut the data, they yielded no support for the belief that people in the Great Lakes basin were dying of cancer more than people elsewhere in the United States and Canada. Surprisingly, the rates for some cancers were actually lower in the Great Lakes area than in some other regions. There was simply no evidence in the public health records of elevated or unusual cancer patterns among those living near the lakes."[21]

A Focus on Cancer Masks the Problem

Colborn was baffled and returned to look over those boxes of wildlife studies again. It is one of those important moments in science when letting go of old theories or preoccupations is the path to follow, guided by instinct, as puzzling new data emerges. There was no clear link to human cancers, and only the most heavily contaminated fish showed cancer tumors. "Yet a long list of fish and animals all across the Great Lakes basin showed ill health that seemed to be impairing their survival." The problem, Colborn wrote later, was the focus solely on cancer. "The phrase is automatic: 'cancer-causing chemical.' The habit of mind is so ingrained; we do not even recognize the conceptual equation that has dominated our thinking about chemicals." The phrase "toxic chemical" was equated with cancer in the minds of the public, regulators, and scientists as well.[22]

Theo Colborn would need to go beyond where Rachel Carson had left off, her life shortened by cancer. Once Colborn began to look at studies with fresh eyes, freed from the focus on cancer, and with the humility and openness of a woman still new to the field, the patterns finally opened up. She got all the files of Canadian researchers Gilbertson and Fox, met with them, and queried them at length, listening and learning as she went. She recalled the presentation in Toronto about abnormal testes in male fish and followed other clues. But her knowledge of current endocrinology was still limited, even with a recent PhD. The field had changed dramatically since her days in undergraduate pharmacy, and ecologists were not trained in endocrinology. She tried various texts, but most were dense, full of acronyms and technical terms, almost unreadable. Finally she found a usable, practical text, *Clinical Endocrine Physiology*, and kept it at her side. But linking animal and human studies remained a problem, too. Most of the human studies had focused on cancer. Finally she recalled a single study out of the thousands she had read. "She dug it out of her files and read it again." Sandra and Joseph Jacobson of Wayne State University had studied women who had regularly eaten Great Lakes fish. They found that their children had been born early, weighed less, and later had neurological problems such as performing poorly on tests of short-term memory that tend to predict IQ.[23]

Colborn assembled all her data, both animal and human, and laid it all in spread sheets. The patterns and exposures she was looking at finally made things clear. The common missing factor was that toxic chemicals were affecting both Herring Gulls and humans through changes they produced in the endocrine system.

Fred vom Saal was one of these who had done critical work on the mechanisms of endocrine disruption that had helped Theo Colborn complete her puzzle. Over the years, vom Saal had shown how differing behavior in laboratory mice was not genetic. The prevailing wisdom was that almost everything was ultimately controlled by genes. The trick was to somehow locate which was the controlling one. With lab mice that were bred from a single original strain, they were genetically virtually identical and only males, it was believed, would show behavior that was aggressive. But vom Saal, through careful observation, discovered tough, aggressive female mice and set out to find out why. "Maybe, vom Saal thought, the answer to the mystery of how genetically identical females could be so different lay in hormones—chemical messengers that travel in the bloodstream, carrying messages from one part of the body to another."[24] Fred vom Saal ultimately revealed that in the typical six-mouse litter, those female fetuses that were positioned in the womb between two males got more *in utero* hormone exposure and showed behavior closest to that of males. The hormones, including testosterone, fluctuated depending on exposure to chemicals. This, in turn, affected how the mice later developed. The results also depended on the timing of exposure, not the size. This also negated the prevailing theory that the dose makes the poison—that there is a linear relationship that shows the higher the dosage, the more toxic the effect. In the case of endocrine disruption, the results, if graphed, tended to follow a U-shape. Tiny doses produced adverse effects, which declined as the dose increased, and then shot up again with large, acute dosages. Another thing that would make all of this difficult for the average person or public official to understand was the incredibly minute exposures that could cause gender and reproductive health effects. The changes that vom Saal observed and reported on were measured not in typical parts per million, or even billion, but in parts per trillion. "One can begin to imagine a quantity so infinitesimally small by thinking of a drop of gin in a train of tank cars full of tonic. One drop in 660 tank cars would be one part in a trillion; such a train would be six miles long." Using this gin and tonic analogy, Colborn describes the male-like characteristics produced in vom Saal's female baby mice as being caused by a cocktail that "had 135 drops of gin in one thousand tank cars of tonic."[25]

Among other kinds of evidence that Colborn needed, tracked down, and assembled to make the case for endocrine disruption compelling is the precise biological mechanism that allows a synthetic chemical to fool the body's system so it will produce hormones when it shouldn't. She describes how she came to understand this through the work and stunningly simple explanations of John McLachlan, a research scientist at the National Institutes for Environment and Health in North Carolina. He had been studying the role of diethylstilbestrol (DES), a synthetic form of the hormone estrogen, for decades and found how synthetic hormones interacted with the estrogen receptor, which normally has a hormone attach to it to create a chemical message sent throughout the endocrine system. Estrogen receptors are special proteins found in cells in the uterus, breasts, brain, liver, and other parts of the body. Colborn describes McLachlan as a born teacher who "has a theatrical flair and a penchant for metaphor." He shows her a large plastic model of estradiol and links it up to the estrogen receptor. "This isn't simply science, it is a fascinating story—the tale of the estrogen receptor, which consorts with so readily with foreigners that it has earned a reputation. Some scientists call it 'promiscuous.' "[26] The artificial estrogenic compounds, it turns out, easily fit into the receptor. Of a number of estrogenic chemicals, those that fit most easily slip in and produce the adverse health effect.

But the science and the story of endocrine disruption, now widely accepted today, needed one more link and one more narrative to make its case persuasive and to vividly portray its dangers. Tiny exposures in lab mice, lesbian Herring Gulls in the Great Lakes, seal die-offs in the North Sea, and mothers who ate lots of locally contaminated fish were one thing. But, how to explain that both polar bears and Inuit mothers, and their children, in the distant, undeveloped Arctic were somehow affected, too? Or that ordinary Americans could be harmed even if they did not live near some contaminated site? To follow the chain of events that leads from a chemical discharge to adverse health effects in a human fetus or baby far away requires imagination, empathy, and knowledge of ecology—the interrelatedness of organisms, ecosystems, and academic disciplines. How can an oily industrial substance like PCBs get all around the globe? It's relatively easy to see how discharges into the Great Lakes or from GE plants near the Hudson River can contaminate local bodies of water and their sediments, why Chicagoans and New Yorkers should beware. But recent tests for chemical "body burdens" show these and other dangerous chemicals are widespread in human blood and tissue around the globe. To show how this can be and how ecology really works, Theo Colborn, Diane Dumanoski, and Pete Myers put at the center of *Our Stolen Future* a saga, an almost incredible, but memorable story of how a single molecule of a PCB could find its way, over many years, from Texas to the distant, frozen Arctic.

Tracking a PCB Molecule from Texas to the Arctic

PCBs were developed in 1929 and moved into widespread production by Monsanto Chemical after it absorbed the Swann Chemical Company in 1935. Nonflammable, PCBs were soon required in transformers and also widely used as lubricants and hydraulic fluids. "They made wood and plastics non-flammable. They preserved and protected rubber. They made stucco waterproof. They became ingredients in paints, varnishes, inks and pesticides. In retrospect, it is clear that their very characteristics that made them a runaway commercial success also made them one of our most serious environmental pollutants."[27] Some reports of toxic effects in workers had surfaced as early as 1936, but manufacturers kept coming up with new, ingenious uses. From 1957 to 1971, for example, before the advent of the Xerox machine, PCBs were put into paper to create carbonless copy paper that was widely used by typists nationwide. By 1964, Sören Jensen, a researcher at the Institute of Analytical Chemistry in Sweden, found a mysterious chemical everywhere he looked when trying to find traces of DDT. He found it throughout the Swedish environment, in the seas, in hair samples he took from his wife and infant daughter, even in wildlife samples that had been collected in 1935 before DDT and other chlorine-based pesticides were in widespread use. After two years, he figured out that the ubiquitous contaminant was PCBs and published the results in the *New Scientist* in 1966. PCBs were finally banned in 1976, but 3.4 billion pounds had been produced. Most of it remains in use today and PCBs are still permitted in closed applications such as transistors and small appliances.

Enter our PCB molecule. No one can know exactly how particular PCBs make it to the Arctic or exactly from where. It is why Theo Colborn and her partners created a fictionalized story, based on documented releases and scientific understanding, to show how molecules are actually transported. There are many kinds (polymers) of PCBs; the one they trace is called PCB-153. Its journey begins in 1947, as the use of PCBs is expanding rapidly after World War II. The Monsanto factory is in Anniston, Alabama, where workers produced a batch in 1947 under the brand name of Aroclor-1254. A half century later, PCB-153 could be found "in the sperm of a man at a fertility clinic in upstate New York, in the finest caviar, in the fat of a newborn baby in Michigan, in penguins in Antarctica, in the Bluefin tuna served at a sushi bar in Tokyo, in the monsoon rains falling in Calcutta, in the milk of a nursing mother in France, in the blubber of a sperm whale cruising the South Pacific, in a wheel of ripe Brie cheese, in a handsome striped bass landed off Martha's Vineyard on a summer weekend. Like most persistent synthetic chemicals, PCBs are world travelers."[28]

The PCB-153 in the Aroclor-1254 begins its journey aboard a freight train headed to Pittsfield, Massachusetts, where it goes to the General Electric plant that makes Pyranol, its own custom formula for transformers that uses Aroclor and other oils. From there, sealed tightly in transformers, it is shipped back by train to the west Texas city of Big Spring, where a transformer containing PCB-153 is installed in a new refinery complex of a small oil and chemical company. Before long, an August thunderstorm sends lighting into the power lines supplying the refinery. The power surge bursts and ruins the transformer, which is then replaced and sent off to the dump. A maintenance worker lifts and tips the broken transformer which spills the oil with PCB-153 onto the dusty parking lot. The dust is blown to Tarzan, Texas, where a housewife sweeps it up and puts it in the trash. Two years later a flash flood sweeps some of the trash and PCBs into a stream. The trash stays put on the stream bank, is parched by the hot Texas sun, and takes to the air as dust again. The dust is carried north by the wind where it is deposited after rain for a brief rest in St. Louis on a bluff beside the Mississippi. Our PCB-153 molecule is then blown north again through Chicago and out over Lake Michigan, where the air cools and the molecule is deposited in the water along with falling rain near Racine, Wisconsin. The PCB-135 attaches itself to algae, avoids being eaten by fish, is impervious to bacteria, and drifts to the bottom of the lake, where it is covered by sediment and stays for years. PCBs, Colborn's story makes clear, are decidedly persistent. PCBs are part of a group of POPs, or persistent organic pollutants, that include DDT, chlordane, aldrin, dieldrin, endrin, toxaphene, heptachlor, and dioxin, which will become the subject of international treaty efforts. As Colborn reminds us, "In *Silent Spring*, they are at the top of Rachel Carson's most-wanted list." But, she explains, "It didn't occur to her to include compounds such as PCBs that may not be particularly poisonous (in the usual sense of causing immediate death or cancer) but are persistent—a fact scientists did not recognize until 1966, four years after *Silent Spring* was published."[29]

Given its persistence, our molecule is again set free in 1956 during work to construct a new waterfront park along the Racine lakefront called Pershing Park. As boulders are dumped along the shore, a huge one hurtles into the sediment and releases PCB-153 once again. It is soon lodged in the fat of a water flea, which is eaten by a tiny shrimplike creature called a mysid; it is eaten by a smelt, which is, in turn, eaten by a trout. Five years later, the trout is caught and packed on ice in a cooler by a vacationing fisherman who drives home toward upstate New York. His car breaks down briefly in the heat and by the time the family reaches home, the trout is spoiled and dumped into the trash. It ends up in a landfill near Rochester, where it is eaten by Herring Gulls. But their eggs are no longer viable and are left on the nest to rot. From there the yolk goes from a skunk that drops it on the shore of Lake Ontario, to a crayfish eaten by an eel that then migrates to the Sargasso Sea off the island of Bermuda. The eel spawns, dies, rots and the molecule, now attached to seaweed, once again moves on and floats north toward Iceland.

There it is eaten by a codepod, another small, shrimplike creature. The codepod is devoured by a cod, which becomes a meal for a ringed seal in the eastern part of the Greenland Sea. The seal is now eaten by a mother polar bear on the Svalbard Islands between Greenland and Norway. She hibernates and then gives birth. The PCB-153 has been stored in her fat and now flows through her breast milk into her hungry cubs. Polar cubs nurse for more than two years and grow to four hundred pounds on a rich diet of their mother's milk. By this time, as PCB-153 has moved along the food chain from predator to predator it has been biomagnified three billion times, joining with other molecules from long ago when it started out in Alabama. Then a decade later, one of the cubs will reach maturity but fail to reproduce.[30] Our PCB molecule has reached the Arctic. It has shown up in the studies of reproductive health problems among polar bears in Svalbard. But given the persistence of PCBs, this is not the only possible destination for our molecule.

Both polar bears and human beings are top predators. As Sandra Steingraber puts it, a human infant drinking contaminated breast milk is at the very top of the food chain, where concentrations of PCBs and other POPs are biomagnified another twenty-seven times more than in the fatty tissues of her mother. Colborn and company explain that PCBs have reached mothers and nursing infants in the Arctic as well as polar bears. Canadian researchers and health officials are concerned about Inuit children on Broughton Island, with a village of 450 people, off Baffin Island, west of Greenland, "more than sixteen hundred miles from the smokestacks of southern Ontario, and twenty-four hundred miles from the industrial centers in Europe." There they have found chronic ear infections and abnormalities in the children's immune systems, including a failure to produce antibodies when they are vaccinated for smallpox, measles, or polio. As Colborn says, "No matter where we live, we share their fate to some degree." Many chemicals that threaten the next generation have found their way into our bodies. There is no safe, uncontaminated place.[31]

The Ethics of Endocrine Disruption: Even Plastic Bottles Can Leach Chemicals

Theo Colborn and her coauthors manage to make even the most arcane of disputes over falling sperm counts interesting and clear. And Colborn actually makes individual scientific studies and their meaning as accessible as an action film. Take the work of Ana Soto and Carlos Sonnenschein, two physician researchers who are part of the relatively small group of alert and attentive scientists whose work forms the basis of our current understanding of endocrine disruption. The threat of hormone-disrupting chemicals has come to light, Colborn says, through accidental discoveries and surprises, "none more bizarre than the incident that began just after Christmas in 1987 at Tufts Medical School in Boston." Soto and Sonnenschein had been working there for more than two decades studying and trying to identify what might inhibit cell growth so that aging cells did not keep growing and run amok, turning into cancer. They had found out that such an inhibitor did exist and were carefully adding estrogen removed from blood serum into estrogen-sensitive breast cancer cells in petri dishes in their lab. Under absolutely strict sterile routines, serum with and without estrogen was added. One set of cancer cells proliferated; the other did not. But in the final week of 1987, something seemed terribly amiss. Both cultures— with and without the estrogen were multiplying like mad. How could their scrupulously meticulous work have suddenly proved wrong? They repeated the experiment again and again, changing only small portions of equipment at a time. Perhaps there was contamination somehow in the tubing, the pipettes, or other things that came in contact with the sera. Finally, after two more years of work, they isolated the contamination to the plastic centrifuges they used that were produced by Corning. They alerted the company, but were told that the composition of the plastic was a "trade secret." More work was now needed to isolate the chemical itself. Finally after more painstaking work, they determined that it was p-nonylphenol, a chemical added to plastics like polystyrene and PVCs to make them more stable and less breakable.[32]

What Soto and Sonnenschein had discovered, and Colborn put into context and made well known, was that endocrine-disrupting chemicals could also leach out from plastics and other seemingly stable substances and then harm human health. Other substances were studied and more was revealed. Baby bottles, plastic children's toys, and more could now hold potential danger.[33] As Colborn makes clear, endocrine-disrupting chemicals harm human health in a variety of ways, are widespread throughout the globe, and often are part of consumer products that once had been seen as surely safe. But she and *Our Stolen Future* ultimately raise larger questions, as did Rachel Carson. Some endocrine-disrupting chemicals have now been removed from consumer products, but most chemicals have not been tested. More are always being produced. And huge amounts of substances like PCBs remain in equipment around the globe. If these threaten human health and reproduction are there not much larger questions of ethics here?

Rachel Carson wrote of human *hubris*, the characteristic the ancient Greeks named as overweening pride, the desire to act like the gods themselves. Colborn says the same. Humans have always risked the unknown, she says, courting both success and catastrophe. But, "our activities no longer involve just one village and its neighbor, one valley or the next. The scale of human activity means that these activities engage the planet." The issues, of course, are complicated and involve a host of value judgments. "It is not just a question of the quality of the science describing the problem but also of how we see the risks and how much risk we are willing to entertain. If all that is at stake is the survival of a single gull colony, it may be wise to wait for further scientific study before embarking on an effort to reduce exposure. If, on the other hand, it is a question of decreasing human sperm counts, prudence may dictate acting immediately rather than waiting to see whether the downward trend continues." There are no glib solutions or recommendations for such dilemmas, Colborn says, or for answering such questions such as whether to stop manufacturing and releasing synthetic chemicals altogether. But, the time has come "to pause and finally ask the ethical questions that have been overlooked in the headlong rush of the twentieth century. Is it right to change Earth's atmosphere? Is it right to alter the chemical environment in the womb for every unborn child?"[34]

Theo Colborn's book, written with Dumanoski and Myers, would have been troubling enough with its compelling, popular narrative of the dangers of widely used chemicals that caused previously little noticed and unconnected health effects on the endocrine systems of animals and humans. That tiny, almost immeasurable, exposures could be the problem, that the outcomes went beyond cancer—to threaten reproductive systems and, possibly, the viability of species only made things worse. So did the ethical questions about synthetic chemicals themselves, the *hubris* of modern science, and the faith of society in unlimited progress.

The Environmental Campaign and the Launch of *Our Stolen Future*

But *Our Stolen Future* also had immediate policy and political implications. Rachel Carson's *Silent Spring* was skillfully launched in Washington, DC, with a large and growing network of environmental organizations, activists, academics, researchers, and political and publishing figures deeply involved in advance, with the plans to launch a public fight over DDT. So it was with the strategy to get the science and the stories of endocrine disruption widely before the American public through the creation and distribution of a book for general readers, *Our Stolen Future*. The collaboration and coauthorship of Theo Colborn, Pete Myers, and Dianne Dumanoski was almost perfect. Colborn had developed a database of thousands of studies over the years and knew the science inside and out. She was chatty, friendly, and disarming, a sort of scientific All-American girl. Myers understood the science and presented a formidable presence, about as far from stereotypes of Birkenstock-clad environmentalists as you can get. And, more important, Myers understood the broader implications of Colborn's work and how to orchestrate a complex political and media strategy. As head of a major environmental foundation, he also had friends and colleagues in numerous environmental organizations, on Capitol Hill, and in the Clinton administration. Dianne Dumanoski, the prize-winning reporter and science writer, brought the journalistic and writing skills to turn complex science into compelling, yet credible stories.

Myers, and colleagues like Phil Clapp of the National Environmental Trust, had access to Vice President Al Gore and had arranged for advance copies of *Our Stolen Future* to be given to him, to the president himself, and to others in the Clinton administration. It was they who arranged for Gore to write a foreword to enhance the book's visibility and importance to policymakers. The vice president more than delivered. He specifically linked *Our Stolen Future* in importance to Rachel Carson's *Silent Spring*. Though Gore was prudent in his pronouncements, he noted that the National Science Foundation had established an expert panel to assess the threat. He called for more research and acknowledged that "we can never construct a society that is completely free of risk." But, the vice president added, the American people have a right to know what substances they are being exposed to and everything that science can tell us about their hazards. *Our Stolen Future* was part of the legacy of Rachel Carson. "Last year I wrote a foreword to the thirtieth anniversary edition of Rachel Carson's classic work, *Silent Spring*. Little did I realize that I would soon be writing a foreword to a book that is in many respects its sequel. . . . *Our Stolen Future* raises questions as profound as those Carson raised thirty years ago. . . . *Our Stolen Future* takes up where Carson left off."[35] Gore's enthusiastic backing was a replay of the blurb for *Silent Spring* by Justice William O. Douglas, who had called Rachel Carson the new Harriet Beecher Stowe, or of President John F. Kennedy's press conference remark that he was aware of the problem of pesticides thanks to Miss Carson and that his administration would look into it. Rachel Carson was immediately faced with an attack campaign orchestrated by the Manufacturing Chemists Association and its corporate allies like DuPont, Monsanto, Dow, and W. R. Grace. Publishers were threatened with lawsuits; public forums were created with doctors and scientists willing to attack Carson. Monsanto even published a parody study called "The Desolate Year." And the campaign had some success. The *New York Times* was initially critical of Carson's work. But, in many ways, the assaults by industry only called more attention to the problem of pesticides that Carson had revealed and that they hoped to deny.

As they had in 1962, industry, its PR men, and its political allies went berserk even before *Our Stolen Future* hit the bookstores in 1996. The attacks were eerily familiar. The *Wall Street Journal* started off by calling the book, and praise for it, an environmental "hype machine" driven by Fenton Communications, an activist-leaning Washington PR-firm. Elizabeth Whelan of the American Council for Science and Health (ACSH) assailed the book as "innuendo piled on top of hypothesis on top of theory." The ACSH is backed by the food, drug, and chemical industries. Whelan got hold of an advance copy and prepared an eleven-page report assailing Colborn's work just before publication. Additional attacks came from the conservative Competitive Enterprise Institute. The Advancement of Sound Science Coalition, which operates out of the offices of APCO Associates, a PR firm that specializes in "grassroots" campaigns for corporate clients, called a press conference with ten scientific skeptics, five of whom were on the board of Elizabeth Whelan's coalition. Concern among manufacturers had long been brewing, and memos leaked to the Sierra Club revealed that well before publication, Gaylee Morgan of John Adams Associates, another PR firm, warned about the impact the book would have, when published, on women and activist coalitions and the attention it would gain. "Up to this point, the general media have not shown much interest in this issue [endocrine disruption and human health]. Because of Colborn's book, however, this may change." Morgan further recommended special efforts by the pesticide and chlorine industries, to use female spokespersons, to create fact sheets for women, and to engage in media outreach to women's publications.[36] But, once more, the attacks only heightened interest in the book and the new danger of endocrine-disrupting chemicals.

Attempts to discredit Theo Colborn, the science of endocrine disruption, and environmentalists in general reappeared and intensified in 2001, when the U.S. Public Health Service announced its conclusion that Steven Arnold, a graduate student in John McLachlan's lab at Tulane University, had committed scientific fraud. McLachlan's studies were among many underpinning Colborn's conclusions, and the announcement specifically cleared McLachlan of any wrongdoing. His lab assistant Arnold had faked data, and then denied it, for a particular study of the heightened, synergistic effects of various endocrine-disrupting chemicals when their exposures are combined. Colborn had indeed referenced this study (among many) in *Our Stolen Future*. The right wing pounced. Two chemical industry supporters, Dennis Avery and Steven Milloy, wrote very similar screeds. There had been questions raised about Arnold's results from the outset, and *Our Stolen Future* had noted in its second edition, which came out shortly after the Arnold paper was published, that the Arnold study could have profound implications, but *only* if its results could be replicated. McLachlan himself tried and failed to replicate such strong synergistic effects in the lab and withdrew Arnold's results in a letter to *Science* magazine, which had originally published the study.

Yet typical of the misrepresentations and overblown rhetoric in the Dennis Avery article was the headline "A Bioterrorist Caught—But Not Punished." Avery claimed that Arnold had asserted that the U.S. food supply is dangerously contaminated, even though he never made such a statement. Avery also claimed that the article caused panic and sent federal authorities off on "a costly wild goose chase." Yet concerns and research had been building among many respected scientists for some time. Avery also claimed, without basis in fact, that the article and fears about endocrine disruption have "cost the United States economy billions of dollars, with the toll still mounting. And the punishment for such bioterrorism is a mere five-year ban from any Federal scientific grants."[37] To top this all off, Avery concluded that the National Research Council reported in 1999 that *no* chemicals in the environment were causing endocrine disruption. He wrote, "In August 1999, an expert committee of the National Research Council—a panel that included representatives of the activist community as well as mainstream scientists—reported there was no evidence that chemicals in the environment were disrupting hormonal process in humans or wildlife." In his rebuttal to this inaccurate assertion, Colborn's coauthor, Pete Myers, states, "This is a fraudulent misrepresentation of the NRC's results." Myers goes on to quote the NRC directly. "There is strong evidence from studies of wildlife and laboratory animals that chemicals can interfere with the body's natural hormone system and disrupt the biological process of development in the womb." The NRC report actually says that high exposures *can* affect human development and that PCBs can do so with moderate exposures, though its authors could not agree about health effects of lower dose exposures, called for more study, and said the subject was still open. Furthermore, the NRC panel had strong industry representation and stopped reviewing evidence after 1997, when many more studies and even stronger evidence supporting Colborn's thesis emerged.[38]

Now in her mid-eighties, with numerous honors, many of them named for Rachel Carson, and following years as a professor and then professor emeritus of zoology at the University of Florida, a center of endocrine and wildlife research, Theo Colborn is still active. She is now, like Sandra Steingraber and others, deeply engaged in the battle against hydraulic fracturing, or "fracking" for natural gas, which pollutes groundwater and worsens global warming. As we talked, she even described her forthcoming article for the peer-reviewed *Environmental Health Studies*. It shows that the environmental damage caused by the release of gases and contaminants during the entire process of extracting natural gas from shale is even worse than that focused on by environmentalists and activists opposed to fracking. Fracking, she says, is only one small part of a much wider set of drilling and extraction practices. Colborn travels less in her eighties and watches her health carefully while running her nonprofit organization, The Endocrine Disruption Exchange (TEDX). TEDX gets its funding from private donors, including a very major one, a wealthy Colorado environmentalist woman, who is glad to see independent research and advocacy that is willing to go beyond the EPA. Theo Colborn's growing think tank is housed, of course, in western Colorado.[39]

Given attacks like those on Colborn and renewed and continuing ones on Rachel Carson herself, it is clear that no single author, no matter how skillful or well received, can shape environmental consciousness or regulation on her or his own. The environmental movement and its closely related environmental health arm have existed since well before Rachel Carson's time and have continued to grow and mature ever since. Ellen Swallow Richards, Florence Merriam Bailey, Alice Hamilton, and other women were all talented organizers, as well as authors and scientists. As the threats to the environment and to human health became more serious and more global in the late twentieth century, women responded with powerful writing and deep commitment, but also with ambitious and effective environmental organizing and leadership as well. Without Rachel Carson and her sisters, we would not have the widespread appreciation for nature and human health that we do today, nor would we have a powerful environmental movement that is often all that prevents even more rapid destruction of the natural and human environment than we have seen. American history and American studies properly laud the contributions of a number of male environmental heroes and leaders like Theodore Roosevelt. But TR would not be carved in the solid, soaring cliffs of Mount Rushmore without the contributions of the many women we have met and many, many others whom we have not. Rachel Carson, of course, deserves her hard-earned place in the pantheon of American environmentalists. She is an extraordinary naturalist, writer, environmental health scientist, and advocate. You and I live tremendously in her debt. But like Harriet Beecher Stowe before her, she is not, and has never been, alone.

EPILOGUE

I F RACHEL CARSON were alive today, she would be astonished at the size and power of the environmental movement she helped to reignite and transform. Women continue to be central to its growth and achievements, even if still underappreciated and underrepresented at the apex of major environmental organizations, academia, and government. When President Barack Obama was first elected, along with a Democratic Congress in 2008, an African American woman, Lisa Jackson, was picked to head the Environmental Protection Agency. Hillary Clinton, a strong environmentalist, became the nation's second woman to sit in Thomas Jefferson's seat as secretary of state. Another woman, Nancy Pelosi, became the first female Speaker of the House of Representatives. An ardent environmentalist, Pelosi's senior staff included veteran environmental women from the National Resources Defense Council, Physicians for Social Responsibility, and other organizations. For the first time in American history, the chair of the Senate Environment and Public Works Committee was also a woman, Barbara Boxer. She held the first serious hearings on global warming in more than a decade. Carol Browner, who had been the longest-serving EPA administrator, headed a new interagency group to coordinate numerous new initiatives concerning clean energy and global warming.

These gains were set back by the off-year election of 2010 when a minority of Americans voted in an ultraconservative, antienvironmental, male-led House of Representatives. Progress on environmental policy, as in the Clinton years, was stymied by a determined minority. And the deepest economic recession since the Great Depression turned all eyes to jobs at any cost, including an intensified search for fossil fuels.

Nevertheless, since *Silent Spring*, the environmental movement has continued to grow from perhaps a dozen nature-oriented national organizations, scattered state and local councils, and a network of scientists, writers, and civic leaders into the largest reform movement in American history. Just in Washington, there are about thirty-four national organizations with some twelve million dues-paying members, millions more electronic activists, and local chapters in every state who cooperate on legislative strategy, organizing, media, and grassroots work. The Internet has allowed further growth of grassroots and regional groups linked to global networks that can mount demonstrations and advocacy campaigns worldwide. The movement's influence is growing once again, especially as the economy improves and there are unmistakable signs of global climate change, including record droughts, fires, and superstorms like Hurricane Sandy. Intense battles over controversial oil and gas extraction methods like fracking and deep-water drilling are spreading.

And women are at the center. Frances Beinecke, president of the Natural Resources Defense Council, has been named by *Fortune* magazine as one of the five most influential women on the American scene. She runs an organization with more than a million members, with hundreds of staff, and a budget in the hundreds of millions. Maggie Fox, former longtime deputy director of the Sierra Club, now heads a vast coalition called Climate Reality, originally established as the Alliance for Climate Protection, launched by Al Gore and the heads of organizations ranging from the environmental Green Group to those concerned with civil rights, business, faith, and much more. It has some thirty million Americans involved with its campaigns. Marjorie Alt heads Environment America, a wide, grassroots network that grew out of the U.S. Public Interest Group, founded by Ralph Nader. Lois Gibbs runs the Center for Health, Environment, and Justice that has trained and activated more than ten thousand local organizations since she helped launch the modern phase of the citizen's environmental health movement after Love Canal in the 1970s. Jamie Rappaport Clark headed the U.S. Fish and Wildlife Service during the Clinton administration, where Rachel Carson was once a junior staffer. She now runs the venerable Defenders of Wildlife. Heather White is the executive director of the Environmental Working Group, with some 1.3 million supporters. Kathleen Rogers is the head of a growing global network of activists for the Earth Day Network (EDN), the organization that continues the legacy of Earth Day started more than forty years ago by Senator Gaylord Nelson (D-WI) and Denis Hayes, now with the Bullitt Foundation. Rogers's EDN has organized worldwide events and, in the United States, among other actions, coordinated more than twelve thousand sermons in churches for Earth Day Sunday. Carol Werner leads the Environmental and Energy Study Institute, the successor to a bipartisan congressional caucus, eliminated by Republicans, that advocates for renewables and clean energy. Thu Pham is the executive director of Rachel's Network, a group of powerful and influential female foundation officials and funders concerned with the environment.

When Rachel Carson was alive, the environmental movement had not yet learned to participate effectively in actual politics, the crafty and sometimes compromised world of real elections. Deb Callahan, now head of the Colorado River Conservation Campaign, has been the longest-serving and first woman to head the League of Conservation Voters (LCV). She built it into one of the strongest political action groups in the nation. Under Callahan, the LCV came agonizingly close to delivering the White House in 2000 and again in 2004 to environmental champions Al Gore and John Kerry. (Kerry's wife, Teresa Heinz Kerry, runs an environmental foundation and think tank and would have been the strongest environmental first lady ever—nosing out Eleanor Roosevelt and Hillary Clinton.)

It is why it now looks, if environmentalists focus on politics and organize for the 2014 elections, that there could soon be an alignment of forces similar to those that moved the American environment onto the national agenda during the Progressive Era of Teddy Roosevelt, the New Deal of Franklin Roosevelt, and the New Frontier of John F. Kennedy. The environmental movement and its allies in policy circles in the Obama administration (many of whom were from the embattled environmental years of Bill Clinton and Al Gore, when antienvironmentalists held sway in one or both chambers of Congress for almost the entire eight years) are also backed up by a host of civic, academic, labor, and religious groups. The scope and numbers would have astonished Ellen Swallow Richards, Alice Hamilton, and other early environmental reformers. A female Episcopal canon (an impossibility when Rachel Carson's funeral was held in the National Cathedral fifty years ago), the Right Reverend Sally Bingham, heads a nationwide network of more than fifteen thousand congregations arrayed in the environmental advocacy group, Interfaith Power and Light. More than 650 college and university presidents, many of them women, have signed the

U.S. College and University Presidents Commitment on Climate Change. It binds their campuses to achieving carbon neutrality, sustainability, and green curricula. During the time Rachel Carson worked on *Silent Spring* and when she died on April 14, 1964, Smith College, where Florence Merriam Bailey began environmental campus organizing, was still run by the last of a long line of male presidents, Thomas Mendenhall, who served from 1959 until 1975.

On campuses there are now large programs and departments in environmental science and environmental studies, sustainability staff, green buildings, and much more. In fact, the transformation of campuses into environmentally astute and active communities was also begun by women. Julian Keniry, a latter-day Florence Merriam Bailey, launched the first modern sustainability efforts in 1989 by organizing the one hundred campus–member Campus Ecology program for the National Wildlife Federation. Women activist academics like Peggy Bartlett of Emory University and Deborah Rowe of Oakland Community College helped start the earliest college-based environmental curriculum and reform efforts. Leading women scholars like Carolyn Merchant of the University of California at Berkeley, and the many others in my sources, have begun the long process of repairing how we see the role of women in shaping our views of the environment, our history, and our government. And women now head American medical schools and schools of public health. They have been elected president of organizations like the American Public Health Association, where Ellen Swallow Richards once angered her mostly male colleagues.

As feminist environmental historian Glenda Riley has said, quoting the early humorist Josh Billings, "Wimmin is everywhere!" So, too, is serious environmental, health, and consumer writing, and advocacy throughout the United States. Why, then, the lack of sufficient progress in meeting the major, dangerous trends toward environmental destruction and harm to human health from resource depletion, toxic chemicals, and global climate change that haunt our era? The answers are similar to, if not the same, as in the days of the first great wave of conservation and environmental health reforms in the Progressive Era. Building a movement to challenge entrenched interests and polluters is difficult, painstaking work. The harm from pollution, poor working conditions, and a plethora of consumer products is so varied, diffuse, and hard to document that it gets lost in the shouting over the price of gasoline or the rise and fall of taxes. It has taken more than a century to build the American environmental movement that we have. It is a vital part of our democracy and of plain old good citizenship. The temptation to place blame on its shortcomings and strategies is strong. But our problems lie with those who seek short-term profit and power at the expense of the rest of us. More writing, more educating, more organizing, more environmental politics, more participation by Americans of all walks of life is what we need. We badly need as well a fundamental love of nature, wonder at its beauties, and a deeply felt and imagined sense of humanity's place within the intricately evolved biological world.

Those who pollute and plunder have huge resources at their command. They challenge serious science, real reform, and claim to care about people and the planet even as they block every reasonable effort to build a better, healthier environment for our children and generations yet to come. Nonetheless, their sway is slowly, steadily, being reduced over time by the determination of ordinary citizens like you and me. We can draw inspiration and leadership from the long line of American women who somehow defied the cinched circumstances and enervated expectations for their gender to become extraordinary leaders of many kinds. They have brought us thus far. We simply need to carry on. Or, like our foremothers, whatever our own stage in life, we can start now down the path that they have set before us.

NOTES

CHAPTER 1. HAVE YOU SEEN THE ROBINS? RACHEL CARSON'S MOTHER AND THE TRADITION OF WOMEN NATURALISTS

1. Quoted in Linda Lear, *Rachel Carson: Witness for Nature* (New York: Henry Holt, 1997), 16–17. Reprinted in 2009 by Houghton Mifflin Harcourt.

2. Lear, *Rachel Carson*, 11–16.

3. Ibid., 13–14.

4. Susan Fenimore Cooper, *Rural Hours* (Syracuse, NY: Syracuse University Press, 1995). All quotations are from this edition.

5. David Jones, "Introduction," in *Rural Hours*, xxxvii.

6. Ibid., xxiii, xi n. 14.

7. Susan Fenimore Cooper, ed., *The Rhyme and Reason of Country Life; or, Selections from the Field Old and New* (New York: G. P. Putnam, 1854). For Susan Cooper and her family's time in Paris, see David McCullough, *The Greater Journey: Americans in Paris* (New York: Simon & Schuster, 2011), 5, 72–74, 94–97. For details of Morse's painting, with identifications, see the National Gallery of Art exhibition, *A New Look: Samuel F. B. Morse, Gallery of the Louvre*, http://www.nga.gov/exhibitions/2011/morse/morseinfo.pdf.

8. Marcia Myers Bonta, *Women in the Field: America's Pioneering Women Naturalists* (College Station: Texas A&M University Press, 1991), 3–4. *Rural Hours* was not reprinted until 1968 (and again in 1995 with a new introduction) by Syracuse University Press. A reedited, scholarly edition followed the Syracuse version in 1998; see Susan Fenimore Cooper, *Rural Hours*, ed. Rochelle Johnson and Daniel Patterson (Athens: University of Georgia Press, 1998). For her nature essays, see Susan Fenimore Cooper, *Essays on Nature and Landscape*, ed. Rochelle Johnson and Daniel Patterson (Athens: University of Georgia Press, 2002).

9. Cooper, *Rural Hours*, 152.

10. Jones, "Introduction," *Rural Hours*, xv.

11. Cooper, *Rural Hours*, 133.

12. Ibid., 133–134.

13. Ibid., 134.

14. Ibid., 135.

15. Ibid., 152–153.

16. Ibid., 154.

17. Ibid., 158.

18. Ibid.

19. Ibid., 174.

20. Ibid., 79.

21. Ibid., 5.

22. Rachel Carson, *Silent Spring* (Boston: Houghton Mifflin, 1962), 105.

23. Madelyn Holmes, *American Women Conservationists: Twelve Profiles* (Jefferson, NC: McFarland, 2004), 8. For "Otsego Leaves," see Cooper, *Essays on Landscape and Nature*, 78–109. For a full list of Susan Cooper's publications, see Rochelle Johnson and Daniel Patterson, eds., *Susan Fenimore Cooper: New Essays on "Rural Hours" and Other Works* (Athens: University of Georgia Press, 2001), 267–270.

24. Holmes, *American Women Conservationists*, 8. For "A Lament for the Birds," see Cooper, *Essays on Landscape and Nature*, 110–114. On Maria Carson and *St. Nicholas* magazine, see Lear, *Rachel Carson*, 18.

25. Graceanna Lewis, *Natural History of Birds: Lectures on Ornithology, Part I* (Philadelphia, 1868).

26. Deborah Jean Warner, *Graceanna Lewis: Scientist and Humanitarian* (Washington, DC: Smithsonian Institution Press, 1979), 103–104, 106.

27. Bonta, *Women in the Field*, 18–20.

28. Ibid., 20–22. Also see Scott Weidensaul, *Of a Feather: A Brief History of American Birding* (New York: Harcourt, 2007), 91–92.

29. Bonta, *Women in the Field*, 24.

30. Ibid.

31. Warner, *Graceanna Lewis*, 81–82, 117–118; and Bonta, *Women in the Field*, 24–26.

32. Maxine Benson, *Martha Maxwell: Rocky Mountain Naturalist* (Lincoln: University of Nebraska Press, 1986), 108.

33. Bonta, *Women in the Field*, 31–36.

34. Quoted in Benson, *Martha Maxwell*, 112.

35. Quoted in Glenda Riley, *Women and Nature: Saving the "Wild" West* (Lincoln: University of Nebraska Press, 1999), 55.

36. Bonta, *Women in the Field*, 30–31, 37–41. Also see Weidensaul, *Of a Feather*, 124–126. A number of photos and stereopticon cards also follow the epilogue in Benson, *Martha Maxwell*, n.p.

37. See Robert Clarke, *Ellen Swallow: The Woman Who Founded Ecology* (Chicago: Follett, 1973); and Caroline E. Hunt, *The Life of Ellen H. Richards, 1842–1911*, anniversary ed. (Washington, DC: American Home Economics Association, 1958); originally published in 1912.

38. Benson, *Martha Maxwell*, 172.

39. Bonta, *Women in the Field*, 40–41.

40. Harriet Kofalk, *No Woman Tenderfoot: Florence Merriam Bailey, Pioneer Naturalist* (College Station: Texas A&M University Press, 1989), 3–11.

41. *A Review of the Birds of Connecticut with Remarks on Their Habits, Etc.* (New Haven, CT: Tuttle, Morehouse and Taylor, 1877); and Kofalk, *No Woman Tenderfoot*, 14–15.

42. Kofalk, *No Woman Tenderfoot*, 20–22.

43. Ibid., 28–30.

44. Ibid., 31.

45. Ibid., 37.

46. John Burroughs, *Writings of John Burroughs*, vol. 1, *Wake-Robin* (Boston: Houghton Mifflin, 1904).

47. Quoted in Kofalk, *No Woman Tenderfoot*, 39.

48. Quoted in ibid., 42.

49. Kofalk, *No Woman Tenderfoot*, 43.

50. Florence A. Merriam, *Birds through an Opera Glass* (Boston: Houghton Mifflin, 1889).

51. Ibid., v.

52. Ibid., vii.

53. Ibid., vi, 4. For Bailey's field guide setting off the modern birding craze and later field guides, see Weidensaul, *Of a Feather*, 195; Mark V. Barrow Jr., *A Passion for Birds: American Ornithology after Audubon* (Princeton, NJ: Princeton University Press, 1998), 156–157; and Kofalk, *No Woman Tenderfoot*, 109.

54. Florence A. Bailey, *A-Birding on a Bronco* (Boston: Houghton Mifflin, 1896). For Fuertes's role, see Holmes, *American Women Conservationists*, 43; and Kofalk, *No Woman Tenderfoot*, 63–70.

55. Florence A. Bailey, *Birds of Village and Field: A Bird Book for Beginners* (Boston: Houghton Mifflin, 1898). Also see Holmes, *American Women Conservationists*, 44–46; Kofalk, *No Woman Tenderfoot*, 80–81; and Barrow, *A Passion for Birds*.

56. Douglas Brinkley, *The Wilderness Warrior: Theodore Roosevelt and the Crusade for America* (New York: Harper Perennial, 2009), 99, 107, 298, 300, 307, 379, 406, 412–413, 492–493, 808; also see Kofalk, *No Woman Tenderfoot*, 73, 85.

57. Lucy Maynard, *Birds of Washington and Vicinity: Where to Find and How to Know Them*, introduction by Florence A. Merriam (Baltimore: Lord Baltimore Press, 1898). Also see Holmes, *American Women Conservationists*, 46–48; and Kofalk, *No Woman Tenderfoot*, 73, 79–85, 87, 89, 102, 157–159. For the Roosevelt bird list, see Brinkley, *Wilderness Warrior*, 813–814, 896n71.

58. Florence Merriam Bailey, *Handbook of Birds of the Western United States* (Boston: Houghton Mifflin, 1902), seventeen editions, including four revisions, until 1935. Frank Chapman, *Handbook of Birds of Eastern North America* (New York: D. Appleton, 1895). Roger Tory Peterson, *Field Guide to the Birds of Eastern North America* (Boston: Houghton Mifflin, 1939). On praise for Bailey's western guide, see Kofalk, *No Woman Tenderfoot*, 104–105. On the impact of Peterson's guide, see Weidensaul, *Of a Feather*, 206–210.

59. Florence Merriam Bailey, *Birds of New Mexico* (Santa Fe: New Mexico Department of Game and Fish, 1928). Also see Kofalk, *No Woman Tenderfoot*, 155; Holmes, *American Women Conservationists*, 52; and B. Elizabeth Horner and Keir B. Sterling, "Feathers and Feminism in the 'Eighties,'" *Smith Alumnae Quarterly* 66, no. 3 (April 1975).

60. Lear, *Rachel Carson*, 142–146; Kofalk, *No Woman Tenderfoot*, 89, 192; Holmes, *American Women Conservationists*, 55. The article alongside Theodore Roosevelt and Mary Austin is Florence Merriam Bailey, "The Spring Migration of Birds," *St. Nicholas* 27 (May 1900): 644–645.

61. Florence Merriam Bailey, "Mrs. Olive Thorne Miller," *Auk: A Quarterly Journal of Ornithology*, 36, no. 2 (April 1919): 163–169; and Bailey, "Olive Thorne Miller," *Condor* 21 (March 1919): 69–73.

62. Deborah Strom, ed., *Birdwatching with American Women: A Selection of Nature Writings* (New York: W. W. Norton, 1986), 4–5; Kofalk, *No Woman Tenderfoot*, 54–58;

Florence Merriam Bailey, *My Summer in a Mormon Village* (Boston: Houghton and Mifflin, 1894).

63. Olive Thorne Miller, *A Bird-Lover in the West* (Boston: Houghton and Mifflin, 1894).

64. Olive Thorne Miller, *The First Book of Birds* (Boston: Houghton Mifflin, 1899), iii.

65. Ibid., 1–2.

66. Ibid., 3.

67. Felton Gibbons and Deborah Strom, *Neighbors to the Birds: A History of Birdwatching in America* (New York: W. W. Norton, 1988), 129, 181. Also see Strom, *Birdwatching with American Women*, xv–xvi.

68. Strom, *Birdwatching with American Women*, 144–146. Also see Frank Graham Jr., with Carl Buchheister, *The Audubon Ark: A History of the National Audubon Society* (New York: Alfred A. Knopf, 1990), 38; and Gibbons and Strom, *Neighbors to the Birds*, 182–183.

69. Mabel Osgood Wright, *Birdcraft: A Field Book of Two Hundred Song Game, and Water Birds*, 9th ed. (New York: Macmillan, 1936).

70. Mabel Osgood Wright and Elliott Coues, *Citizen Bird: Scenes from Bird-Life in Plain English for Beginners* (New York: Macmillan, 1897), 2.

71. Ibid., 12.

72. Neltje Blanchan, with an introduction by John Burroughs, *Bird Neighbors: An Introductory Acquaintance with One Hundred and Fifty Birds Commonly Found in the Gardens, Meadows, and Woods about Our Homes* (New York: Doubleday & McClure, 1897).

73. Frank Nelson Doubleday, *The Memoirs of a Publisher* (Garden City, NY: Doubleday, 1972).

74. Gibbons and Strom, *Neighbors to the Birds*, 184–185. Also see Bonta, *Women in the Field*, 155, 164.

75. Anna Botsford Comstock, *Handbook of Nature Study*, with a foreword by Verne N. Rockcastle (Ithaca, NY: Cornell University Press, 1986), 1 (emphasis in the original).

76. Anna Botsford Comstock, *The Comstocks of Cornell: John Henry Comstock and Anna Botsford Comstock: An Autobiography by Anna Botsford Comstock* (Ithaca, NY: Comstock Publishing Associates and Cornell University Press, 1953).

77. Mary Hunter Austin, *The Land of Little Rain* (Boston: Houghton Mifflin, 1903); and *Earth Horizon* (New York: Literary Guild, 1932).

78. Lear, *Rachel Carson*, 138.

79. Glenda Riley, *Women and Nature*.

80. Quoted in Lear, *Rachel Carson*, 132.

81. Quoted in Paul Brooks, *Speaking for Nature: How Literary Naturalists from Henry Thoreau to Rachel Carson Have Shaped America* (San Francisco: Sierra Club Books, 1980), 184–186. Also see Esther F. Lanigan, *Mary Austin: Song of a Maverick* (Tucson: University of Arizona Press, 1997), 23–24, on Austin's views of English and science and how a stimulating environment overcame earlier depression.

82. Quoted in Riley, *Women and Nature*, 62–63.

83. Brooks (quoting Austin), *Speaking for Nature*, 187, 189–190.

84. Mary Austin, "Scavengers," in Deborah Strom, ed., *Birdwatching with American Women: A Selection of Nature Writings* (New York: W. W. Norton, 1986), 163–164.

85. Ibid., 166–167.

86. Ibid., 168.

87. Brooks, *Speaking for Nature*, 191–192; and Strom, *Birdwatching*, 161–162.

88. The quotation and other information from Austin's autobiography in this paragraph are from Brooks, *Speaking for Nature*, which contains no footnotes. Austin on the Boulder Dam is quoted in Riley, *Women and Nature*, 64.

89. Brooks, *Speaking for Nature*, 191.

90. Ibid., 192.

91. Quoted in Graham, *Audubon Ark*, 39–40.

92. Riley, *Women and Nature*, 1.

93. Quoted in Carolyn Merchant, ed., *Major Problems in American Environmental History: Documents and Essays*, 2nd ed. (Boston: Houghton Mifflin, 2005), 342–343.

94. Merchant, *Major Problems*, 343–346.

CHAPTER 2. DON'T HARM THE PEOPLE: ELLEN SWALLOW RICHARDS, DR. ALICE HAMILTON, AND THEIR HEIRS TAKE ON POLLUTING INDUSTRIES

1. Mabel Osgood Wright, *Birdcraft*, 9th ed. (New York: Macmillan, 1936); Susan Fenimore Cooper, *Rural Hours* (Syracuse, NY: Syracuse University Press, 1995); and Susan Fenimore Cooper, "Otsego Leaves I: Birds Then and Now," in *Susan Fenimore Cooper: Essays on Nature and Landscape*, ed. Rochelle Johnson and Daniel Patterson (Athens: University of Georgia Press), 80–81.

2. Quoted in Paul Brooks, *The House of Life: Rachel Carson at Work* (Boston: Houghton Mifflin, 1972), 232.

3. Linda Lear, *Rachel Carson: Witness for Nature* (New York: Henry Holt, 1997), 315.

4. Ibid.

5. Robert Clarke, *Ellen Swallow: The Woman Who Founded Ecology* (Chicago: Follett, 1973), 108–109, 133, 146.

6. Lear, *Rachel Carson*, 67–70.

7. Clarke, *Ellen Swallow*, 5–13; Caroline L. Hunt, *The Life of Ellen H. Richards: 1842–1911* anniversary ed. (Washington, DC: American Home Economics Association, 1958), 4–21; Mary Joy Breton, *Women Pioneers in the Environment* (Boston: Northeastern University Press, 1998), 47–48.

8. Clarke, *Ellen Swallow*, 17–19; Hunt, *The Life of Ellen H. Richards*, 23, 35; Breton, *Women Pioneers*, 48.

9. Clarke, *Ellen Swallow*, 21–23; Hunt, *The Life of Ellen H. Richards*, 32, 37–40; Breton, *Women Pioneers*, 48–49.

10. Clarke, *Ellen Swallow*, 23–24; Hunt, *The Life of Ellen H. Richards*, 40–42; Breton, *Women Pioneers*, 49.

11. Clarke, *Ellen Swallow*, 25, 38–39; Hunt, *The Life of Ellen H. Richards*, 49–50; Breton, *Women Pioneers*, 49.

12. Robert Hallowell Richards, *His Mark* (Boston: Little, Brown, 1936), 152–153.

13. Quoted in Clarke, *Ellen Swallow*, 46.

14. Ibid., 48–49.

15. Ibid., 52, for the quotation; and Hunt, *The Life of Ellen H. Richards*, 70.

16. Clarke, *Ellen Swallow*, 52–53; Hunt, *The Life of Ellen H. Richards*, 71.

17. Clarke, *Ellen Swallow*, 97–98. For Ellen Swallow Richards's pioneering role and her teaching and mentoring of the first true American university engineering dean, see "Purdue Engineer's Link to Ellen Swallow Richards," Purdue University College of Consumer and Family Sciences, http://www.cfs.purdue.edu/about/history/ellen_richards/engineering.html.

18. Clarke, *Ellen Swallow*, 100–101.

19. Quoted in ibid., 82.

20. Clarke, *Ellen Swallow*, 85; Hunt, *The Life of Ellen H. Richards*, 111–112.

21. Quoted in Clarke, *Ellen Swallow*, 89.

22. Quoted in ibid., 91–93.

23. Clarke, *Ellen Swallow*, 95–96.

24. Ibid., 101–104.

25. Quoted in ibid., 110–111.

26. Clarke, *Ellen Swallow*, 111.

27. Ibid., 144–149.

28. Richards and Monroe quoted in ibid., 160–162.

29. Quoted in Clarke, *Ellen Swallow*, 163.

30. Clarke, *Ellen Swallow*, 166–167; Hunt, *The Life of Ellen H. Richards*, 154–156.

31. Clarke, *Ellen Swallow*, 177–178; Hunt, *The Life of Ellen H. Richards*, 154.

32. Clarke, *Ellen Swallow*, 233, 236; Hunt, *The Life of Ellen H. Richards*, 173, 176; Breton, *Women Pioneers*, 61–62.

33. Madeleine P. Grant, *Alice Hamilton: Pioneer Doctor in Industrial Medicine* (London: Abelard-Schumann, 1967), 67; Rachel Carson, *Silent Spring* (Boston: Houghton Mifflin, 1962), 332; Alice Hamilton and Harriet L. Hardy, *Industrial Toxicology*, 2nd ed. (New York, Hoeber, 1949); and "Memorial Minute for Harriet L. Hardy," *Harvard Faculty Gazette*, May 1, 1997, http://www.news.harvard.edu/gazette/1997/05.01/FacultyofMedici.html.

34. Edith Hamilton, *The Greek Way* (New York: W. W. Norton, 1964). For Alice Hamilton on Edith, see Alice Hamilton, *Exploring the Dangerous Trades: The Autobiography of Alice Hamilton, M.D.* (Boston: Little, Brown, 1943), 18, 19, 37, 44, and 360. On Miss Porter's, see Barbara Sicherman, *Alice Hamilton: A Life in Letters* (Cambridge, MA: Harvard University Press, 1984), 22; and Hamilton, *Exploring the Dangerous Trades*, 35–37.

35. Hamilton, *Exploring the Dangerous Trades*, 36–38; Grant, *Alice Hamilton*, 36; Robert Gottlieb, *Forcing the Spring: The Transformation of the American Environmental Movement* (Washington, DC: Island Press, 1993), 47.

36. Hamilton, *Exploring the Dangerous Trades*, 38; Grant, *Alice Hamilton*, 36; Sicherman, *A Life in Letters*, 35.

37. Hamilton, *Exploring the Dangerous Trades*, 38–34; for fuller treatment of Hamilton at the University of Michigan Medical School, see Grant, *Alice Hamilton*, 36–45. Also, Sicherman, *A Life in Letters*, 35–37.

38. Hamilton, *Exploring the Dangerous Trades*, 42–45. Also see Grant, *Alice Hamilton*, 45–50.

39. Hamilton, *Exploring the Dangerous Trades*, 51.

40. Ibid., 53.

41. Ibid., 54

42. Clarke, *Ellen Swallow*, 136–137; Hunt, *The Life of Ellen H. Richards*, introduction to 6th ed., n.p.; Marcia Myers Bonta, *Women in the Field: America's Pioneering Women Naturalists* (College Station: Texas A&M University Press, 1991), 189; Harriet Kofalk, *No Woman Tenderfoot: Florence Merriam Bailey, Pioneer Naturalist* (College Station: Texas A&M University Press, 1989), 43–44, 53–55.

43. Grant, *Alice Hamilton*, 70–72; Hamilton, *Exploring the Dangerous Trades*, 115–118; Gottlieb, *Forcing the Spring*, 48–51.

44. Quoted in Grant, *Alice Hamilton*, 74–75.

45. Grant, *Alice Hamilton*, 75.

46. Steven Johnson, *The Ghost Map: The Story of London's Most Terrifying Epidemic— and How It Changed Science, Cities, and the Modern World* (New York: Riverhead Books, 2006). Johnson's is the best version of this often recounted classic public health tale.

47. Grant, *Alice Hamilton*, 78–79.

48. Ibid., 85.

49. Grant, *Alice Hamilton*, 87–88; and Sicherman, *A Life in Letters*, 158–159 and 166.

50. Grant, *Alice Hamilton*, 52; Sicherman, *A Life in Letters*, 158.

51. Grant, *Alice Hamilton*, 97–98. Alice Hamilton outlived her sister, dying, at age 101, on September 22, 1970. Also see Sicherman, *A Life in Letters*, 416.

52. Grant, *Alice Hamilton*, 98.

53. Ibid.

54. Ibid., 100.

55. Grant, *Alice Hamilton*, 136–140; Sicherman, *A Life in Letters*, 237–238; Hamilton, *Exploring the Dangerous Trades*, 252–253.

56. Grant, *Alice Hamilton*, 154–155; Alice Hamilton, *Industrial Poisons in the United States* (New York: Macmillan, 1925). The notice that pleased Hamilton most was praise from Dr. (Sir) Thomas Oliver, author of the *Dangerous Trades*. Also see Sicherman, *A Life in Letters*, 240.

57. Devra Davis, *The Secret History of the War on Cancer* (New York: Basic Books, 2007), 380–381.

58. Ibid., 382.

59. Ibid., 384.

60. Robert K. Musil, *Hope for a Heated Planet: What Americans Are Doing to Fight Global Warming and Build a Better Future* (New Brunswick, NJ: Rutgers University Press, 2009), 69–70.

61. Grant, *Alice Hamilton*, 147–152; Hamilton, *Exploring the Dangerous Trades*, 286–288.

62. Harriet L. Hardy, *Challenging Man-Made Disease: The Memoirs of Harriet L. Hardy, M.D.* (New York: Praeger, 1983), 65–66.

63. Hardy, *Challenging Man-Made Disease*, 66–67; and "Memorial Minute for Harriet L. Hardy," *Harvard Faculty Gazette*, May 1, 1997.

64. Hardy, *Challenging Man-Made Disease*, 175. Also see Boston Women's Health Book Collective, *Our Bodies, Ourselves* (Boston: New England Free Press, 1971). This edition was based on a 1970 course edition in stapled newsprint put out by New England Free Press. The 1971 edition sold 250,000 copies, mostly by word of mouth. The first commercial edition, followed by many others, is from Simon & Schuster in 1973.

65. Hardy, *Challenging Man-Made Disease*, 177–178.

66. Ibid., 182–183.

67. Carson, *Silent Spring*, 338; Hardy, *Challenging Man-Made Disease*, 69.

68. Devra Davis, *When Smoke Ran Like Water: Tales of Environmental Deception and the Battle against Pollution* (New York: Basic Books, 2002), 66–67; Daniel Costa and Terry Gordon, "Mary O. Amdur," *Toxicological Sciences* 56, no. 1 (2000): 5–7; http://toxsci .oxfordjournals.org/content/56/1/5.full.

69. Davis, *When Smoke Ran Like Water*, 68–70; Costa and Gordon, "Mary O. Amdur," 5.

70. Quoted in Davis, *When Smoke Ran Like Water*, 74–77; Costa and Gordon, "Mary O. Amdur," 5.

CHAPTER 3. CARSON AND HER SISTERS:
RACHEL CARSON DID NOT ACT ALONE

1. Linda Lear, *Rachel Carson: Witness for Nature* (New York: Henry Holt, 1997), 208, 226, 273, 336, 378, 398, 426.

2. Ibid., 409, 413, 417.

3. Ibid., 17–20, 23–25.

4. Gene Porter-Stratton, *Freckles* (New York: Grosset & Dunlap, 1904); and *A Girl of the Limberlost* (New York: Grosset & Dunlap, 1909). I reference throughout a later edition reprinted from the original by Triangle Books in 1944. For an online version of *A Girl of the Limberlost*, see Project Gutenberg, http://www.gutenberg.org/ebooks/ 125 or, http://www.online-literature.com/stratton-porter/girl-of-the-limberlost/2/. For excerpts from *Freckles*, see Deborah Strom, ed., *Birdwatching with American Women: A Selection of Nature Writings* (New York: W. W. Norton, 1986), 67–81; and commentary in Paul Brooks, *Speaking for Nature: How Literary Naturalists from Henry Thoreau to Rachel Carson Have Shaped America* (San Francisco: Sierra Club Books, 1980), 177–180. Also see Vera Norwood, *Made from This Earth: American Women and Nature* (Chapel Hill: University of North Carolina Press, 1993), 74–77.

5. Ralph Lutts, *The Nature-Fakers* (Golden, CO: Fulcrum Publishing, 1990), 25–27.

6. Rachel Carson, *Silent Spring* (Boston: Houghton Mifflin, 1962), 127.

7. Gene Stratton-Porter, *A Girl of the Limberlost* (New York: Triangle Books, 1944), 207.

8. Ibid., 261.

9. Quoted in Judith Reick Long, *Gene Stratton-Porter: Novelist and Naturalist* (Indianapolis: Indiana Historical Society, 1990), 238. Also see the Gene Stratton-Porter Historical Site website, "Gene Stratton-Porter & Her Limberlost Swamp," http://www .genestrattonporter.net.

10. John Elder, "Withered Sedge and Yellow Wood: Poetry in *Silent Spring*," in Peter Matthiessen, ed., *Courage for the Earth: Writers, Scientists, and Activists Celebrate the Life and Writing of Rachel Carson* (Boston: Houghton Mifflin, 2007), 79–95.

11. Quoted in Lear, *Rachel Carson*, 32–35.

12. Ibid., 47.

13. Robert Clarke, *Ellen Swallow: The Woman Who Founded Ecology* (Chicago: Follett, 1973), 147.

14. Lear, *Rachel Carson*, 47, 58–61.

15. Ibid., 68–69.

16. Ibid., 69–70.

17. Ibid., 72–73.

18. For women science faculty in the area at the time, see Margaret Rossiter, *Women Scientists of America: Struggles and Strategies to 1940*, vol. 1 (Baltimore: Johns Hopkins University Press, 1982), 207–208. For a short biography of Pearl by Rachel Carson's genetics professor at Johns Hopkins, see H. S. Jennings, "Raymond Pearl, 1879–1940," in *Biographical Memoirs. National Academy of Sciences* 22 (1943): 295–347. Also see "Pearl, Raymond," *Complete Dictionary of Scientific Biography*, 2008, Encyclopedia.com, http://www.encyclopedia.com/doc/1G2-2830903322.html; accessed October 2, 2011.

19. Michael Purdy, "Occupational Health's Dynamo," *Johns Hopkins Public Health Magazine*, Fall 2001, http://www.jhsph.edu/magazineFall01/Prologues.htm.

20. Quoted in ibid.

21. Ibid.

22. Anna Baetjer, *Women in Industry: Their Health and Efficiency* (Philadelphia: W. B. Saunders, 1946).

23. Ibid., 309.

24. Purdy, "Occupational Health's Dynamo."

25. Lear, *Rachel Carson*, 78–79.

26. Ibid., 124, 132, 138.

27. Shirley Ann Briggs, *A Basic Guide to Pesticides: Their Characteristics and Hazards* (Washington, DC: Taylor and Francis, 1992); and "Distinguished Alumni Award, Shirley A. Briggs, for Achievement, 1995," University of Iowa Alumni Association, http://www.iowaalum.com/daa/briggs.html.

28. Lear, *Rachel Carson*, 126–127. Carson, *Silent Spring*, 119–120. For Rosalie Edge, an important New Deal–era environmentalist, see Dyana Z. Furmansky, *Rosalie Edge: Hawk of Mercy, the Activist Who Saved Nature from the Conservationists*, foreword by Bill McKibben, afterword by Roland C. Clement (Athens: University of Georgia Press, 2010).

29. Lear, *Rachel Carson*, 126, 139.

30. Ibid., 181–182.

31. Scott Weidensaul, *Of a Feather: A Brief History of American Birding* (New York: Harcourt, 2008), 206–215; and Elizabeth J. Rosenthal, *Birdwatcher: The Life of Roger Tory Peterson* (Guilford, CT: Lyons Press, 2008), 27–29, 51, 73–74.

32. Lear, *Rachel Carson*, 118–120.

33. Ibid., 118–119.

34. Ibid., 120–121; and William Souder, *On a Farther Shore: The Life and Legacy of Rachel Carson* (New York: Crown Publishers, 2012), 136, 154.

35. Lear, *Rachel Carson*, 91, 105.

36. Ibid., 120–121, 155. Also see Tom Hilchey, "Mary Sears, 92, Oceanographic Editor and Scientist at Woods Hole," *New York Times*, September, 10, 1997; and Military Sea-lift Command, Ship Namesakes "T-AGS 65 USNS Mary Sears," http://www.msc.navy.mil/inventory/citations/marysears.htm. Also see Hali Felt, *The Story of the Remarkable Woman Who Mapped the Ocean Floor* (New York: Henry Holt, 2013) about Marie Thorp, who drew the first maps of the ocean bottom.

37. Lear, *Rachel Carson*, 154, 169. For background on *Stride toward Freedom*, see Martin Luther King Jr. and the Global Freedom Struggle, http://mlk-kpp01.stanford.edu/index.php/encyclopedia/encyclopedia/enc_stride_toward_freedom_the_montgomery_story_1958. On Marie Rodell, King, and the FBI, see The Martin Luther King, Jr. Papers Project, http://mlk-kpp01.stanford.edu/primarydocuments/Vo14/13-Nov-1957_ToRodell.pdf; and The Martin Luther King, Jr. FBI File, Part II: The King-Levison File, http://academic.lexisnexis.com/documents/upa_cis/10734_MLKJrFBIFilePt2.pdf. For Rachel Carson's FBI file, see Living on Earth: PRI's Environmental News Magazine, "Rachel Carson Remembered," May 25, 2007, http://www.loe.org/shows/segments.html?programID=07-P13-00021&segmentID=1. Also see Souder, *On a Farther Shore*, 344.

38. Lear, *Rachel Carson*, 171–172.

39. Paul Brooks, *The House of Life: Rachel Carson at Work* (Boston: Houghton Mifflin, 1972), xi; and Brooks, *Speaking for Nature*, 276, 285.

40. Lear, *Rachel Carson*, 280, 301–302.

41. Ibid., 127. Ada Clapham Govan, *Wings at My Window* (New York: Macmillan, 1940).

42. See for example, Neltje Blanchan, *Birds Every Child Should Know* (New York: Grosset & Dunlap, 1907); or Mabel Osgood Wright and Elliott Coues, *Citizen Bird: Scenes from Bird-Life* (New York: Macmillan, 1897). On Wright, see Strom, *Birdwatching with American Women*, 144–146.

43. Lear, *Rachel Carson*, 129.

44. Govan, *Wings at My Window*, 187.

45. Olive Thorne Miller, *The First Book of Birds* (Boston: Houghton, Mifflin, 1899), 2.

46. Blanchan, *Birds Every Child Should Know*, vii–viii.

47. Lear, *Rachel Carson*, 127–129, 283. Also see Rachel Carson, *The Sense of Wonder*, introduction by Linda Lear, photos by Nick Kelsh (New York: Harper, 1998).

48. Lear, *Rachel Carson*, 389.

49. Rosa Parks, with Jim Haskins, *Rosa Parks: My Story* (New York: Dial Books, 1992), 101–116; Douglas Brinkley, *Rosa Parks* (New York: Viking Penguin, 2000), 54–59, 67–77, 90–96; Taylor Branch, *Parting of the Waters: America in the King Years, 1951–63* (New York: Simon and Schuster, 1988), 120–123. Also see Robert K. Musil, *Hope for a Heated Planet: What Americans Are Doing to Fight Global Warming and Build a Better Future* (New Brunswick, NJ: Rutgers University Press, 2009), 82–83, 111–112, for a broad view of the civil rights movement. For Huckins and her correspondence with Beatrice

Hunter of the Committee against Mass Poisoning, see Lear, *Rachel Carson*, 313–314. William Souder does not mention Olga Huckins in *On a Farther Shore*, offering instead a more traditional view of Carson as relatively "apolitical" and downplaying her connections to the environmental movement and to women. Compare with Mark Lytle on the importance of women and networks in Mark Hamilton Lytle, *The Gentle Subversive: Rachel Carson, "Silent Spring," and the Rise of the Environmental Movement* (Oxford: Oxford University Press, 2007), 7–8, 246–247.

50. Lear, *Rachel Carson*, 318, 415.

51. Lawrence S. Wittner, *Resisting the Bomb: A History of the World Disarmament Movement, 1954–1970* (Stanford, CA: Stanford University Press, 1997), 363. Also see Milton S. Katz, *Ban the Bomb: A History of SANE, the Committee for a SANE Nuclear Policy, 1957–1985* (Westport, CT: Greenwood Press, 1986).

52. Wittner, *Resisting the Bomb*, 132–134.

53. Ibid., 319. Also see "Marjorie Spock, Rachel Carson, Eurythmy," Waldorf in the Home: Resources for Nourishing Family Life, http://www.waldorfinthehome.org/2008/02/marjorie_spock_rachel_carson_e.html. William Souder downplays Marjorie Spock and emphasizes her most eccentric qualities. Souder, *On a Farther Shore*, 275, 284, 286, 288–290, 303, 309.

54. Lear, *Rachel Carson*, 320.

55. Ibid. Also see Norwood, *Made from This Earth*, 153–162, for women's networks, including local activists.

56. Lear, *Rachel Carson*, 321.

57. Ibid., 322.

58. Quoted in ibid., 332.

59. Joshua Blu Buhs, *The Fire Ant Wars: Nature, Science, and Pollution Policy in the Twentieth-Century* (Chicago: University of Chicago Press, 2004), 127–136.

60. Quoted in Lear, *Rachel Carson*, 322.

61. Joseph J. Hickey and L. Barrie Hunt, "Songbird Mortality Following Annual Programs to Control Dutch Elm Disease," *Atlantic Naturalist* 15, no. 2 (1960): 87–92; and Hickey and Hunt, "Some Effects of Insecticides on Terrestrial Birdlife," *Report of the Subcommittee on Relation of Chemicals to Forestry and Wildlife*, State of Wisconsin, January 1961, 2–43. Also see Carson, *Silent Spring*, 107n, 109n, 317–318. Hickey's article with Daniel W. Anderson on global eggshell thinning in *Science* magazine was a major reference for Frank Graham Jr.'s follow-up call to arms, *Since Silent Spring: Rachel Carson Has Been Proved Right, What Have We Done about It?* (Boston: Houghton Mifflin, 1970), 132.

62. Joseph J. Hickey, "Foreword" in Frances Hamerstrom, *Birding with a Purpose: Of Raptors, Gabboons, and Other Creatures* (Ames: Iowa State University Press, 1984), viii.

63. Frances Hamerstrom, *Strictly for the Chickens* (Ames: Iowa State University Press, 1980).

64. Joseph S. Hickey, ed., *Peregrine Falcon Populations: Their Biology and Decline* (Madison: University of Wisconsin Press, 1969), 64; Hamerstrom, *Strictly for the Chickens*, 96, 111, 115, 132–134. Also see Frances Hamerstrom, *My Double Life: Memoirs of a Naturalist* (Madison: University of Wisconsin Press, 1994), 244–246. Stephanie Fabritius (dean and professor of biology, Centre College, KY), interviews with the author, November 11, 2010, and April 16, 2011.

65. Lear, *Rachel Carson*, 387–389. Also see Norwood, *Made from This Earth*, 157.

66. Lear, *Rachel Carson*, 404–406.

67. Ibid., 399–400.

68. Ibid., 342.

69. Ibid., 344.

70. Musil, *Hope for a Heated Planet*, 126–127.

71. Lear, *Rachel Carson*, 399.

72. Ibid., 400. Also see Souder, *On a Farther Shore*, 321. Souder does not mention the role of Agnes Meyer of the *Washington Post* and other women in the DC launch of *Silent Spring*.

73. Lear, *Rachel Carson*, 410–411.

74. All quoted in ibid., 419.

75. Lear, *Rachel Carson*, 414–415.

76. Quoted in ibid., 419.

77. Ibid., 415.

78. Ibid., 452.

CHAPTER 4. RACHEL CARSON, TERRY TEMPEST WILLIAMS, AND ECOLOGICAL EMPATHY

1. Robert K. Musil, *Hope for a Heated Planet: How Americans Are Fighting Global Warming and Building a Better Future* (New Brunswick, NJ: Rutgers University Press, 2009), 5–6. Descriptions of Kennedy's speech are from personal conversations and also interviews with Ted Sorensen by the author for *Consider the Alternatives* radio show, which was syndicated weekly from 1972 to 1992. Its archives are at the Swarthmore College Peace Collection, Swarthmore, PA. On the antinuclear movement and the Limited Test Ban Treaty, see Lawrence Wittner, *The Struggle Against the Bomb*, vol. 2, *Resisting the Bomb: A History of the World Nuclear Disarmament Movement, 1954–1970* (Stanford, CA: Stanford University Press, 1997), ix–x, 246–261, 416–423.

2. Linda Lear, ed., *Lost Woods: The Discovered Writing of Rachel Carson* (Boston: Beacon Press, 1998), 192–196. Lear has collected Carson's introduction, "To Understand Biology," from *Human Biology Projects* (New York: Animal Welfare Institute, 1960) and the "Preface" to the highly influential British book by Ruth Harrison, *Animal Machines: The New Factory Farming Industry* (London: Vincent Stuart, 1964). Also see Vera Norwood, *Made from This Earth: American Women and Nature* (Chapel Hill: University of North Carolina Press, 1993), 161–163, on Carson's long association with the animal rights activist Christine Stevens.

3. Wittner, *Resisting the Bomb*, 2, 8, 146–148, 153–154. For a detailed recounting of the *Lucky Dragon* incident, see William Souder, *On a Farther Shore: The Life and Legacy of Rachel Carson* (New York: Crown Publishers, 2012), 223–227, 229–234.

4. Linda Lear, *Rachel Carson: Witness for Nature* (New York: Henry Holt, 1997), 237. For Operation Crossroads, the 1946 atomic tests at Bikini Atoll, see the Naval History and Heritage Command, http://www.history.navy.mil/faqs/faq76-1.htm. For fuller accounts, see Jonathan Weisgall, *Operation Crossroads: The Atomic Tests at Bikini Atoll* (Annapolis, MD: Naval Institute Press, 1994); and Gregg Herken, *The Winning Weapon: The Atomic Bomb in the Cold War, 1945–1950* (New York: Alfred A. Knopf, 1980).

5. For Operation Capricorn and the work of Roger Revelle in oceanography see Deborah Day, "Roger Randall Dougan Revelle Biography," Scripps Institution of Oceanography Archives, http://www.scilib.ucsd.edu/sio/biogr/Revelle_Biogr.pdf.

6. Lear, *Rachel Carson*, 333; Wittner, *Resisting the Bomb*, 2, 8, 146–148, 153–154.

7. Lear, *Rachel Carson*, 105, 199.

8. Ibid., 524n39, 375.

9. Wittner, *The Struggle against the Bomb*, vol. 1: *One World or None: A History of the World Disarmament Movement through 1953* (Stanford, CA: Stanford University Press, 1993), 66–67; and Norman Cousins, *Modern Man Is Obsolete* (New York: Viking Press, 1946).

10. Rachel Carson, "The Pollution of Our Environment," in Lear, *Lost Woods*, 232; and 240–241 for the Committee on Nuclear Information.

11. Rachel Carson, "Preface to the Second Edition of *The Sea around Us*," in Lear, *Lost Woods*, 101.

12. Ibid., 106–109.

13. Carson, "The Pollution of Our Environment," in Lear, *Lost Woods*, 238–239.

14. Ibid., 240–242.

15. Wittner, *Resisting the Bomb*, 151; Also see Howard Ball, *Justice Downwind: America's Testing Program in the 1950s* (New York: Oxford University Press, 1986), 59, 86, 89, 92, 95, 126, 164.

16. Terry Tempest Williams, "The Moral Courage of Rachel Carson," in Peter Matthiessen, ed., *Courage for the Earth: Writers, Scientists, and Activists Celebrate the Life and Words of Rachel Carson* (Boston: Houghton Mifflin, 2007), 130.

17. Ibid., 145.

18. Ibid., 132.

19. Ibid., 145.

20. Ibid.

21. Lear, *Rachel Carson*, 138, 142, 145. For Carson's Bear River booklet, see the digital archives, "Conservation History," of the U.S. Fish and Wildlife Service, http://training.fws.gov/history/HistoricDocuments.html.

22. Harriet Kofalk, *No Woman Tenderfoot: Florence Merriam Bailey, Pioneer Naturalist* (College Station: Texas A&M University Press, 1989), 57; and Florence Merriam, *My Summer in a Mormon Village* (Boston: Houghton Mifflin, 1894), 163–164. Also available as a Nook e-book and in a reprint edition from Nabu Press.

23. Olive Thorne Miller, *A Bird-Lover in the West* (Boston: Houghton Mifflin), 212.

24. Ibid., 213.

25. Kofalk, *No Woman Tenderfoot*, 59.

26. Quoted in ibid.

27. Ibid., 208; Bailey's *Condor* article is Florence Merriam Bailey, "White-Throated Swifts at Capistrano," *Condor* 9 (November 1907): 169–172. Also see Florence Merriam Bailey, *Among the Birds in the Grand Canyon Country* (Washington, DC: U.S. Department of the Interior, National Park Service, U.S. Government Printing Office, 1939). For Martha Maxwell, see Marcia Myers Bonta, *Women in the Field: America's Pioneering Naturalists* (College Station: Texas A&M University Press, 1991), 32–39; and Maxine Benson, *Martha Maxwell: Rocky Mountain Naturalist* (Lincoln: University of Nebraska Press, 1986), 128–149.

28. On the role of the Utah Federation of Women's Clubs, see Glenda Riley, *Women and Nature: Saving the "Wild" West* (Lincoln: University of Nebraska Press, 1999), 100, 102–103, 164–165. For Williams on Mormon environmentalism, see Michael Austin, ed., *A Voice in the Wilderness: Conversations with Terry Tempest Williams* (Logan: Utah State University Press, 2006), 95–96; and Terry Tempest Williams, William B. Smart, and Gibbs M. Smith, *New Genesis: A Mormon Reader on Land and Community* (Salt Lake City, UT: Gibbs-Smith, 1998).

29. Terry Tempest Williams, "Introduction," to Mary Austin, *The Land of Little Rain* (New York: Penguin Books, 1997), xi–xii.

30. Ibid., ix–x.

31. Terry Tempest Williams, *When Women Were Birds: Fifty-Four Variations on Voice* (New York: Farrar, Straus and Giroux, 2012), 21.

32. Terry Tempest Williams, *An Unspoken Hunger: Stories from the Field* (New York: Vintage, 1994), 13–15.

33. Williams, *When Women Were Birds*, 29–31.

34. Ibid., 62–63.

35. Austin, *A Voice in the Wilderness*, 172.

36. Williams, *When Women Were Birds*, 33–36; and Lear, *Rachel Carson*, 100.

37. Quoted in Austin, *A Voice in the Wilderness*, 15–16; emphasis in the original.

38. Williams, *When Women Were Birds*, 75–80.

39. Austin, *A Voice in the Wilderness*, 79. Terry Tempest Williams, *Between Cattails*, pictures by Peter Parnall (New York: Charles Scribner's Sons, 1985).

40. Austin, *A Voice in the Wilderness*, 4, 8, 64 (for the quotation); and Terry Tempest Williams, *Coyote's Canyon*, photographs by John Telford (Salt Lake City, UT: Gibbs-Smith, 1989).

41. Terry Tempest Williams, *Refuge: An Unnatural History of Family and Place* (New York: Vintage Books, 1991), 15.

42. Ibid., 14.

43. Ibid., 16–17.

44. Ibid., 17.

45. Ibid., 18.

46. Ibid., 22.

47. Ibid., 24.

48. Ibid., 89.

49. Ibid., 89–90.

50. Ibid., 90.

51. Ibid., 178.

52. Ibid., 290. For a detailed listing of the numbers and types of U.S. nuclear weapons tests, see the Federation of American Scientists, http://www.fas.org/nuke/guide/usa/nuclear/nv209nar.pdf.

53. Terry Tempest Williams, *Finding Beauty in a Broken World* (New York: Pantheon Books, 2008), 5.

54. Ibid., 11.

55. Ibid., 23–24.

56. Ibid., 31, 110.

57. Ibid., 24.

58. Ibid., 33–34.

59. Ibid., 44.

60. Ibid., 38.

61. Ibid., 39.

62. Ibid., 45. On the role of C. Hart Merriam and Vernon Bailey in the poisoning of prairie dogs and other "varmint" eradication efforts, see Donald Worster, *Nature's Economy: A History of Ecological Ideas*, 2nd ed. (Cambridge: Cambridge University Press, 1994), 262–265; Richard Manning, *Rewilding the West: Restoration in a Prairie Landscape* (Berkeley: University of California Press, 2009), 122, 138, 140, 169; Susan Jones, "Becoming a Pest: Prairie Dog Ecology and the Human Economy in the Euroamerican West," *Environmental History* 4, no. 4 (October 1999): 531–552; and Douglas Brinkley, *The Wilderness Warrior: Theodore Roosevelt and the Crusade for America* (New York: Harper Perennial, 2009), 411.

63. Williams, *Finding Beauty*, 87.

64. Ibid., 89.

65. Ibid., 93. John L. Hoogland, *The Black-Tailed Prairie Dog: Social Life of Burrowing Animals* (Chicago: University of Chicago Press, 1995).

66. Williams, *Finding Beauty*, 95.

67. Ibid., 102.

68. Ibid., 101.

69. Ibid., 125–126.

70. Ibid., 127.

71. Ibid., 135.

72. Ibid., 139.

73. Ibid., 189.

74. Ibid., 197; emphasis in the original.

75. Ibid., 204.

76. Ibid., 209–212.

77. Ibid., 224.

78. Ibid.

79. Ibid., 225.

80. Ibid.

81. Ibid., 227.

82. Ibid., 260.

83. Ibid., 261; emphasis in the original.

84. Ibid., 361.

85. Carson, *Silent Spring*, 6.

86. Williams, *When Women Were Birds*, 194–197.

87. Stephen Trimble and Terry Tempest Williams, *Testimony: Writers of the West Speak on Behalf of Utah Wilderness* (Minneapolis: Milkweed Editions, 1996).

88. Ibid., 4.

89. Ibid., 10.

90. Ibid., 7.

CHAPTER 5. THE ENVIRONMENT AROUND US AND INSIDE US: ELLEN SWALLOW RICHARDS, *SILENT SPRING*, AND SANDRA STEINGRABER

1. Mary Joy Breton, *Women Pioneers for the Environment* (Boston: Northeastern University Press, 1998), 48. On Richards's early work with Mitchell, also see Robert Clarke, *Ellen Swallow: The Woman Who Founded Ecology* (Chicago: Follett, 1973), 17–19; and Caroline Hunt, *The Life of Ellen H. Richards*, anniversary ed. (Washington, DC: American Home Economics Association, 1958, originally published in 1912), 23, 35.

2. Terry Tempest Williams, *Finding Beauty in a Broken World* (New York: Pantheon Books, 2008), 44.

3. Steven Johnson, *The Ghost Map: The Story of London's Most Terrifying Epidemic—and How It Changed Science, Cities, and the Modern World* (New York: Riverhead Books, 2006), 27–30, 54–55.

4. Sandra Steingraber, "*Silent Spring*: A Father-Daughter Dance," in Peter Matthiessen, ed., *Courage for the Earth: Writers, Scientists, and Activists Celebrate the Life and Writing of Rachel Carson* (New York: Houghton Mifflin, 2007), 56–57, 51–54.

5. Ibid., 55.

6. Ibid., 58.

7. Ibid.

8. Ibid., 59.

9. Steingraber, "A Father-Daughter Dance," 60. Rachel Carson knew Shawn and wrote to him directly after E. B. White declined to write an article. But Carson first came to the attention of Shawn and got a *New Yorker* piece that led to her best-seller, *The Sea around Us*, because a female part-time assistant editor, Edith Oliver, loved her first five chapters so much that she brought them to Shawn's attention. See William Souder, *On a Farther Shore: The Life and Legacy of Rachel Carson* (New York: Crown Publishers, 2012), 142.

10. Steingraber, "A Father-Daughter Dance," 57.

11. Sandra Steingraber, *Living Downstream: An Ecologist Looks at Cancer and the Environment* (New York: Addison Wesley, 1997), 2.

12. Ibid.

13. Ibid.

14. Ibid., 3

15. Ibid., 5.

16. Ibid., 15.

17. Ibid., 16.

18. Ibid., 17.

19. Quoted in ibid.

20. Steingraber, *Living Downstream*, 136–137.

21. Steingraber, *Having Faith: An Ecologist's Journey to Motherhood* (Cambridge, MA: Perseus Publishing, 2001); and Sandra Steingraber, *Raising Elijah: Protecting Our Children in an Age of Environmental Crisis* (Philadelphia: Da Capo Press, 2011).

22. Steingraber, *Having Faith*, ix.

23. Ibid., 37–38.

24. Ibid., 39.

25. Ibid., 43.

26. Ibid., 44 46.

27. Ibid., 75.

28. Ibid., 82.

29. Ibid., 89.

30. Ibid., 90–92.

31. Ibid., 94.

32. Ibid., 96–97.

33. Ibid., 100–101.

34. Ibid., 108.

35. Ibid., 115.

36. Ibid., 115–116.

37. Ibid., 117.

38. Ibid., 120.

39. Ibid., 261.

40. Ibid., 262.

41. Ibid., 286.

42. Steingraber, *Raising Elijah*, xiii.

43. Ibid.

44. Ibid., xi.

45. Ibid., xii.

46. Ibid.

47. Clarke, *Ellen Swallow*, 80–81.

48. Steingraber, *Raising Elijah*, 35–38, 52–55.

49. Ibid., 77.

50. Ibid., 73.

51. Ibid., 272–273. For an audio-visual online essay on fracking, written and read by Steingraber for *Orion* magazine in 2012 in honor of the fiftieth anniversary of *Silent Spring*, see http://www.orionmagazine.org/index.php/audio-video/item/audio_slide_show_the_fracking_of_rachel_carson.

CHAPTER 6. RACHEL CARSON, DEVRA DAVIS, POLLUTION, AND PUBLIC POLICY

1. Devra Davis, *When Smoke Ran Like Water: Tales of Environmental Deception and the Battle against Pollution* (New York: Basic Books, 2002); and Devra Davis, *The Secret History of the War on Cancer* (New York: Basic Books, 2007). For the quotation, Devra Davis, interview with author, September 21, 2012, Washington, DC.

2. Davis, *When Smoke Ran Like Water*, xiii–xv.

3. Ibid., 1–3.

4. Ibid., 3.

5. Ibid., 7.

6. Ibid., 8.

7. Ibid., 9.

8. Ibid., 22–28. For unions and the environment, generally, and the United Steel Workers on Donora, see Chad Montrie, *A People's History of Environmentalism in the United States* (London: Continuum, 2013), 74–75.

9. Davis, *When Smoke Ran Like Water*, 48–53.

10. Ibid., 58–59.

11. Ibid., 86.

12. Ibid., 90–91, 95.

13. Ibid., 100–103.

14. Ibid., 104–109.

15. Ibid., 121.

16. Ibid., 122.

17. Ibid., 153–154.

18. Ibid., 156.

19. Ibid., 158.

20. Ibid., 260; and Robert K. Musil, *Hope for a Heated Planet: How Americans Are Fighting Global Warming and Building a Better Future* (New Brunswick, NJ: Rutgers University Press, 2009), 29.

21. Davis, interview with author, September 21, 2012, Washington, DC.

22. Davis, *When Smoke Ran Like Water*, 160–161.

23. Ibid., 166–167.

24. Ibid., 168–169.

25. Ibid., 172.

26. Ibid., 181–182.

27. Ibid., 186.

28. Ibid., 190.

29. Quoted in ibid., 214–218. The bracketed words are Milloy's paraphrase of Davis; the bracketed single closing quotation mark is my insertion.

30. Davis, *The Secret History of the War on Cancer*, ix–x.

31. Ibid., x–xi.

32. Ibid., 17–19.

33. Ibid., 23.

34. Ibid.

35. Ibid., 74–77.

36. Ibid., 99.

37. Ibid., 100.

38. Ibid., 101–102.

39. Ibid., 103–104.

40. Ibid., 78–79.

41. Ibid., 80–81.

42. Ibid., 82.

43. Ibid., 93.

44. Ibid., 96.

45. Ibid., 363.

46. Ibid., 364.

47. Ibid., 366–367.

48. Ibid., 368–372.

49. Ibid., 376.

50. Ibid., 377–378.

51. Ibid., 382–388.

52. See Devra Davis, *Disconnect: The Truth about Cell Phone Radiation, What the Industry Is Doing to Hide It, and How to Protect Your Family* (New York: Dutton Adult, 2010); and Davis's Environmental Health Trust website, http://www.ehtrust.org.

CHAPTER 7. RACHEL CARSON AND THEO COLBORN:
ENDOCRINE DISRUPTION AND ETHICS

1. For biographical highlights, see "Theodora E. Colborn: A Biography," http://www.bookrags.com/printfriendly/?=bios&u=theodora-e-colborn-wob. Also, Theo Colborn, telephone interview with author, July 27, 2012.

2. Glenda Riley, *Women and Nature: Saving the "Wild" West* (Lincoln: University of Nebraska Press, 1999), 101.

3. Ibid., 109.

4. Ibid., 124–126. For Mary Cronin, as well as interesting discussion of the relationship between women's climbing costumes and the emergence of "The New Woman," see Susan R. Schrepfer, *Nature's Altars: Mountains, Gender, and American Environmentalism* (Lawrence: University of Kansas Press, 2005), 76 ff.

5. Riley, *Women and Nature*, 143–144.

6. Susan Goodman and Carl Dawson, *Mary Austin and the American West* (Berkeley: University of California Press, 2008), 213–214.

7. Linda Lear, *Rachel Carson: Witness for Nature* (New York: Henry Holt, 1997), 180, 258. On women in the Dinosaur National Monument campaign, see Schrepfer, *Nature's Altars*, 210–211.

8. Colborn, interview with author, July 27, 2012.

9. Ibid.

10. Theo Colborn, Dianne Dumanoski, John Peterson Myers, foreword by Vice President Al Gore, *Our Stolen Future: Are We Threatening Our Fertility, Intelligence, and Survival?—A Scientific Detective Story* (New York: Dutton, 1996), 12–13; "Theodora E. Colborn: A Biography"; and Colborn, interview with author, July 27, 2012.

11. Colborn, interview with author, July 27, 2012.

12. Theodora E. Colborn et al., *Great Lakes, Great Legacy?* (Washington, DC: Conservation Foundation and the Institute for Research on Public Policy, 1990); Theo Colborn and Coralie Clement, eds., *Chemically-Induced Alterations in Sexual and Functional Development: The Wildlife/Human Connection*, vol. 21, *Advances in Modern Environmental Toxicology* (Princeton, NJ: Princeton Scientific Publishing Co., 1992); Colborn, interview with author, July 27, 2012. The Wingspread statement can be found in the appendix to *Our Stolen Future*, 252–260.

13. Colborn, interview with author, July 27, 2012.

14. Colborn, *Our Stolen Future*, 1–2.

15. Ibid., 3–4.

16. Ibid., 4–5.

17. Ibid., 6.

18. Ibid., 10.

19. Ibid., 14–15.

20. Ibid., 16–17.

21. Ibid., 18–19.

22. Ibid., 19.

23. Ibid., 20–24.

24. Ibid., 31.

25. Ibid., 40.

26. Ibid., 70–72.

27. Ibid., 89–90.

28. Ibid., 91–92.

29. Ibid., 91–96.

30. Ibid., 97–103.

31. Ibid., 107–108.

32. Ibid., 122, 127–128.

33. Ibid., 122–129.

34. Ibid., 246–247.

35. Ibid., v–vi.

36. David Helvarg, "When Science Fails, Try Public Relations: The Chemical Industry's Attempt to Discredit *Our Stolen Future*," *Sierra* 82, no. 1 (January/February 1997).

37. See an analysis and rebuttal of attacks on *Our Stolen Future* by John Peterson Myers, http://www.our stolenfuture.org/myths/2002–0120avery.htm.

38. Ibid.

39. Colborn, interview with author, July 27, 2012. Also see the TEDX website, http://www.endocrinedisruption.com/home.php.

INDEX

Note: Page numbers in *italics* indicate figures.

AAAS (Association for the Advancement of Science), 63, 87

AAUW (American Association of University Women; earlier, Association of Collegiate Alumnae), 65, 120

A-Birding on a Bronco (Bailey), 37, 38

abolitionist movement, 15, 25, 27, 51, 188, 193

Abzug, Bella, 207, 208, 209, 210

acetaldehyde, 179

acro-osteolysis, 221

ACSH (American Council for Science and Health), 248

Adams, Ansel, 135–136

Addams, Jane, 6, 74–75, 79–80. *See also* Hull House

Advancement of Sound Science Coalition (TASSC), 211, 248

Agassiz, Elizabeth Cabot, 24

Agassiz, Louis, 24

Agassiz Association for Nature Study, 24

Agency for Toxic Substances Disease Registry (ATSDR), 181

Agent Orange, 183

Agricultural Research Service (ARS), 54–55, 90, 114–115, 116

agriculture: DES use in livestock, 180; federal grazing regulations, 50; methyl bromide and methyl iodide use in, 192. *See also* herbicides; organic gardening and farming; pesticides

AHEA (American Home Economics Association), 70

air pollution: deaths from, 198–199, 202, 205–206; industry's resistance to regulations on, 201–202, 203–205; killer smogs, 86, 87, 199–200, 201, 205

air pollution toxicology: Devra Davis's work in, 201–203; founder of, 99–100; health effects and costs studied, 204–205; Mary Amdur's work in, 85–87, 165, 202; techniques, 86, 87, 202

Albatross III (ship), 107

Alldredge, Charles, 106, 118, 119

Allen, Durwood, 114

Allen, Joel A., 50

alligators, 210, 236

Alt, Marjorie, 253

Alzheimer's disease, 167

AMAX Corporation, 232

Amdur, Mary O.: air pollution studies of, 85–87, 165, 202; approach of, 56; industry's attack on, 87–88; lead studies of, 86

American Association for Labor Legislation, 77

American Association for the Advancement of Women, 26, 42

American Association of University Women (AAUW; earlier, Association of Collegiate Alumnae), 65, 120

American Cancer Society, 219

American Chemical Council, 224

American Council for Science and Health (ACSH), 248

American Economics Association, 77

American flamingo (*Phoenicopterus ruber*), 145–146

American Home Economics Association (AHEA), 70

American Indians, 27. *See also* Arctic

American Industrial Hygiene Association, 87–88

American Institute of Mining and Mineralogical Engineers, 63

ABOUT THE AUTHOR

Environmental leader and avid birder Robert K. Musil is the author of *Hope for a Heated Planet: How Americans Are Fighting Global Warming and Building a Better Future* (Rutgers University Press, 2009). He teaches environmental politics and history at American University in Washington, DC, where he is a senior fellow at the Center for Congressional and Presidential Studies. Trained in environmental health, literature, and the humanities, Musil holds degrees from Yale, Northwestern, and the Johns Hopkins School of Public Health and honorary degrees in science and humane letters.

Musil was the longest-serving CEO of the Nobel Peace Prize–winning Physicians for Social Responsibility and is a Woodrow Wilson Visiting Fellow and popular lecturer on college campuses. A former award-winning, nationally syndicated broadcaster, his most recent voicing is as narrator of the prize-winning documentary film *Scarred Lands and Wounded Lives: The Environmental Footprint of War.*